Language and the Brain

Language and the Brain

A SLIM GUIDE TO NEUROLINGUISTICS

Jonathan R. Brennan

OXFORD
UNIVERSITY PRESS

Great Clarendon Street, Oxford, OX2 6DP,
United Kingdom

Oxford University Press is a department of the University of Oxford.
It furthers the University's objective of excellence in research, scholarship,
and education by publishing worldwide. Oxford is a registered trade mark of
Oxford University Press in the UK and in certain other countries.

© Jonathan R. Brennan 2022

The moral rights of the author have been asserted

Impression: 1

All rights reserved. No part of this publication may be reproduced, stored in
a retrieval system, or transmitted, in any form or by any means, without the
prior permission in writing of Oxford University Press, or as expressly permitted
by law, by licence, or under terms agreed with the appropriate reprographics
rights organization. Enquiries concerning reproduction outside the scope of the
above should be sent to the Rights Department, Oxford University Press, at the
address above.

You must not circulate this work in any other form
and you must impose this same condition on any acquirer.

Published in the United States of America by Oxford University Press
198 Madison Avenue, New York, NY 10016, United States of America.

British Library Cataloguing in Publication Data

Data available

Library of Congress Control Number: 2021947523

ISBN 978-0-19-881475-7 (Hbk)
ISBN 978-0-19-881476-4 (Pbk)

DOI: 10.1093/oso/9780198814757.001.0001

Printed and bound by
CPI Group (UK) Ltd, Croydon, CR0 4YY

Links to third-party websites are provided by Oxford in good faith and
for information only. Oxford disclaims any responsibility for the materials
contained in any third-party website referenced in this work.

For Lisa

Contents

Acknowledgments viii
List of figures x

1. Introduction — 1
2. The toolbox — 18
3. Sounds in the brain — 44
4. A neural code for speech — 62
5. Activating words — 82
6. Representing meaning — 100
7. Structure and prediction — 116
8. Composing sentences — 139
9. Building dependencies — 156
10. Wrapping up — 166

Abbreviations 174
Glossary of terms 175
International Phonetic Alphabet for English 178
Notes 179
Figure acknowledgments 186
References 191
Index 204

Acknowledgments

This book is only possible due to the insight, generosity, and support of a great many people. To begin, I owe an indescribable debt to my mentors for their time, energy, and wisdom which goes far beyond teaching me "how to be a cognitive scientist" (they did that too): Liina Pylkkänen, Robin Queen, and Rick Lewis.

As an actual document, rather than some vague idea in the back of my mind, I am grateful for and in awe of the hard work and editorial insights of Julia Steer and her team at OUP.

A wonderful aspect of my little corner of science is the amazing group of people who are working on all things relating to language, cognition, and the brain. I am very lucky to be able to call these people "my colleagues." Special thanks goes to Doug Bemis, Julie Boland, Suzanne Dikker, Dave Embick, John Hale, Matt Husband, Ioulia Kovelman, Ellen Lau, Lisa Levinson, Alec Marantz, Andrea Martin, Lars Meyer, Chistophe Pallier, Colin Phillips, David Poeppel, Hugh Rabagliati, Tim Roberts, Ed Stabler, Jon Sprouse, Sarah Van Wagenen, Ming Xiang, and Eytan Zweig. (Many more names could be added to this already long list.) My friend and office neighbor Sam Epstein in particular had a tremendous influence on my understanding of how (and how not) to link brains and language; I only wish he were here to read the result.

The book developed out of disorganized notes and slides that first surfaced in the classroom. I'm lucky to have been joined there by many spectacular students. If they are a reflection of the future of our field, we're in pretty great shape. Almost every page bears some impression of the conversations from those classes, and from walking down the hall, in the lab, or on the way to get a cup of coffee. In particular, I must acknowledge Jeonghwa Cho, Lauretta Cheng, Justin Craft, Tamarae Hildebrandt, Chiawen Lo, Emily Sabo, Tzu-yun Tung, Neelima Wagley,

Rachel Weissler, and also students at the 2013 and 2017 Linguistic Society of America Summer Schools, and the 2018 Linguistic Society of Korea Winter School.

Finally, there is no way – no way at all – this book would ever have happened without the intellectual and emotional support of my partner, Lisa Levinson, and the truly boundless energy of our three children. Thank you.

List of figures

1. Auditory localization in the barn owl. — 6
2. Two strategies for building structure. — 7
3. Over 150 years of language in the brain. — 11
4. Three views of the author's brain. — 19
5. The central nervous system. — 20
6. The neuron. — 23
7. The deficit/lesion method. — 28
8. The fMRI method. — 32
9. Electrophysiological methods. — 36
10. Auditory pathway. — 45
11. Speech information unfolds on multiple time-scales. — 46
12. The neurogram. — 50
13. Analysis by synthesis. — 60
14. A dual-stream model for speech perception. — 63
15. The neural representation of phonemes. — 65
16. Voxel-based lesion symptom mapping (VLSM). — 84
17. The time-course and localization of lexical activation. — 88
18. Stages of spoken-word recognition. — 93
19. The semantic system. — 103
20. Semantic dementia. — 105
21. Components of sentence structure. — 117
22. The N400 event-related potential. — 121
23. Predictability and the N400. — 125
24. How semantic predictions might unfold. — 130
25. The P600 ERP. — 132
26. How syntactic predictions might unfold. — 135
27. Prediction dynamics. — 137
28. A combinatoric brain network. — 141

29. Simple composition and the anterior temporal lobe. 145
30. Syntax in the posterior temporal lobe. 151
31. Dependency processing in the LIFG. 159

1

Introduction

Humans use language. Other animals don't. Let's go ahead and start there. If you ask someone "why?" they might give the perfectly reasonable answer that humans have human brains, and human brains are equipped to use human language. That's not all so different from observing that I'm never going to be able to find pollen like a honeybee or learn to sing the song of a starling because I'm neither a bee nor a songbird. Seriously, when it comes to birdsong, I'm no good. I lack what the ethologist Peter Marler (1991) called the "instinct to learn" that particular skill. So, what is it about the human brain that make it able to use language?

The pursuit of this question is the focus of intense research across a whole range of scientific disciplines, including language scientists interested in uncovering the mental computations and representations that make language possible, and neuroscientists whose focus is on how brains are wired in order to learn and use information. I'm sorry to say that scientists don't yet have a comprehensive answer to how this all works (well, not too sorry. If we did, I'd be out of a job). What we do have is a firm foundation of careful observations and compelling theories, along with a whole host of avenues of active research. This book introduces you to this exciting field: neurolinguistics.

We'll discuss how the brain transforms waves of sound pressure into meaningful words, how meaning itself is represented by large networks of neurons, and how brain regions work together to make sense of the never-heard-before phrases and sentences that you encounter, and make yourself, every day. These topics, and a whole host of others discussed ahead, are all pieces in the very large puzzle facing neurolinguistics.

I want you to learn how these pieces are starting to fit together. But, if you are like me, you don't work on a puzzle just to see the picture at the end. It's the act of solving the puzzle – even finding how to place one particularly tricky piece – that is satisfying. So along with facts and figures, I'm going to spend quite a bit of time talking about questions. What are the questions that have helped to establish some of the foundations for neurolinguistics? What questions are guiding current research, and where might they be leading us even if we're not there yet?

Here's the thing: Observations, facts, results, or findings – the typical "stuff" of science – are really only as good as the question they try to answer. Ask a bad question and no amount of sophisticated equipment and careful observation will lead to truly deep understanding. Let's, then, consider together what makes a "good question."

Linking neurons to noun phrases

A good question has a recognizable answer. In other words, I think the most useful scientific questions are those for which you can work out what a possible answer looks like. For example, consider a question like:

> How does the brain represent nouns?

What would an answer to this question look like? Is it a map of brain regions involved in understanding and producing nouns? Is it a definition of what counts as a well-formed noun phrase in a particular language, or perhaps a computer algorithm that recognizes nouns and puts them into sentences? Let's say I point you to a linguistics textbook and say something like "See here, nouns are defined like so, and over here are the grammatical rules for how they work in English…". Have I answered your question? I suspect you might not be sure that I have, and you might not be sure what a satisfying answer would even look like.

To help make questions more precise, cognitive scientists tend to distinguish three different types answers, or *levels of description*, that we might be looking for.[1] The first level of description is to define the problem that the brain is trying to solve. In vision, the brain must convert a two-dimensional pattern of light and dark from the retina of your eye into a three-dimensional map of your dining room populated by a table, chairs, your cat, and so forth. In speech, the brain converts continuous sound waves to discrete speech sounds, *phonemes* like /k/, /æ/, and /t/. Then, these speech sounds activate words and their meanings, such as the noun that refers to a small furry animal in my house. Answers of this sort, that indicate the problem that the brain is solving – its inputs and outputs – is called a *computational* description of the brain.

Another level of description targets the steps a system goes through in order to solve a particular computational problem. For example, you might wonder if the brain recognizes speech sounds by keeping a kind of table, one which matches up sound waves with appropriate speech sounds. Then, whenever it hears a sound, it could look up which speech sound "matched" the acoustic waveform. This sort of answer is an *algorithmic* description of the brain – it specifies the steps the brain would need to carry out to solve some particular computational problem.

Lastly, you might consider how neurons carry out the algorithm that solves a problem. How could neurons represent a table of information? How do they interact to "look up" items in the table? This last sort of answer is an *implementational* description of the brain.

Each level of description captures a different kind of question about how the brain works. There is a name for the possible answers we can start to consider when we have a nice precise question: *hypotheses*. Separating hypotheses into levels of description has proven incredibly

Table 1. **Levels of description.** Understanding a cognitive system like language requires formulating questions (and answers) on at least three different levels of description.

Computational	The problem a cognitive system is solving, including the inputs and outputs.
Algorithmic	The steps by which a cognitive system solves the problem, yielding the correct outputs for the provided inputs.
Implementational	How a physical system (such as a neural circuit) carries out a particular algorithm.

useful in asking, and answering, more precise questions. We can see, for example, that each of the answers given above to the question "How does the brain represent nouns?" serves as a different *kind* of answer – each answer addresses a different level of description.

One consequence of looking at language through the lens of these levels of description is that it separates different kinds of research. So, a syntactician may carry on studying the nature of grammatical structures found in human languages (a computational-level question) without worrying much about how neural circuits might interact to support those structures. This is moreorless the same as how a zoologist can uncover the intricate communication system of the honey-bee without ever measuring neuronal firing rates.[2] Moreover, this perspective makes it clear that we can't learn, for example, how the brain represents nouns by simply measuring brain activity and seeing what happens. To answer this kind of question, we would need first to develop some ideas of what a "noun" is (a computational-level description), add to that a hypothesis of how the brain recognizes such objects (algorithm), and then turn our attention to how neurons might work together to implement those ideas.[3]

Let me give a few examples to make these ideas more concrete. The first example considers what happens if we focus just on one level: the implementation. It's about video games. The key idea is to see how well a neuroscientist could make sense of a video game system by only paying attention to electrical circuits. Instead of measuring properties of neurons, these neuroscientists recorded the connections and electrical discharges of a computer microprocessor (Jonas and Kording, 2017). In fact, the researchers approach this challenge in several ways, drawing inspiration from many of the different tools and techniques deployed in the neurosciences (these tools will be the main focus of Chapter 2). For example, one strategy is to disrupt the actions of individual transistors and see how that affects the operation of the whole system. This is somewhat similar to studying the brain through the lens of neurological disorders; we will in fact start to do something similar in the very next section. Another strategy is to record electrical discharges from different parts of the microprocessor and try to correlate those with states of the video game being played.

None of these strategies, nor any of the several others pursued in this study, worked to make sense of the microprocessor. We know this because the operation of that particular microprocessor is, of course, completely understood. It was engineered, after all. So, the researchers could examine the truth of any of the many statistical generalizations that emerged after disrupting transistors, correlating electrical discharges, mapping electrical connections, etc. As they put it: "in the case of the processor we know its function and structure and our results stayed well short of what we would call a satisfying understanding" (Jonas and Kording, 2017, p. 14). The warning message is clear: Scientists will struggle to get a deep understanding of language in the brain by focusing only on its implementation. Research must draw on insights from all of the levels of description.

The second example shows how answers at these different levels can be productively combined. It is about owls. Barn owls use sound to find (and catch) their prey. Research into how the owl does this offers a wonderful illustration of the payoff when care is taken to distinguish different levels of description.[4] The research begins with a computational description of the problem that the owl's brain must solve: Sound waves reaching each ear must be converted into a location representing where the sound comes from. For simplicity, let's just consider the direction of the sound relative to the owl's head. Owls have available to them two symmetrical sound detectors (a.k.a. "ears"). These allow for one of several algorithms that can solve this computation by calculating the difference in timing and loudness of a sound that is detected at each ear.

One way to calculate such differences uses two engineering building blocks: delay lines and coincidence detectors. Imagine two parallel pathways, one beginning at the left ear and one at the right. Each path has forks at fixed intervals. When a signal reaches each ear, it passes down that pathway, and reaches each fork one after another at a fixed rate; this is a delay line. At some point, these two signals, moving in opposite directions, will meet. The forks from each line are connected to a row of detectors. These coincidence detection circuits "light up" if and only if both of their inputs are receiving a signal at the same time. The point along these rows where the coincidence detector lights up – that is, where the signals from both delay lines intersect – is directly proportional to where *in space* the sound originally came from. If the sound comes from the right side of the owl, then it will reach the right ear and beginning traveling down that delay line first. Because the signal reached the right ear first, it travels further along the right delay line as compared to the left. So, the point at which the signals intersect, and the coincidence detector lights up, is further from the right ear (and closer to the left). This kind of circuit is illustrated in Fig. 1. Sets of neurons in a small part of the owl's brain stem called the nucleus laminaris appear to operate precisely in the way just described. This neural circuit serves to implement the algorithm of coincidence detection, which in turn carries out the computation of mapping from sound to a location in space.

Remarkably, and this is really important, other brains convert sounds to locations quite differently. The field mouse, which may very well be the owl's prey in the situation above, also uses sound to estimate where something (such as a swooping owl) might be. But mice use different algorithms and different neuronal implementations, for example based on differences in sound intensity and frequency rather than timing, to solve the same computational challenge of converting sound to location (Grothe et al., 2010).

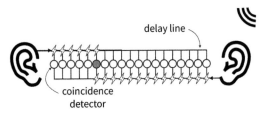

Figure 1. Auditory localization in the barn owl. To find prey, barn owls must convert the sounds heard to locations. One algorithm to achieve this computation combines a pair of delay lines, one from each ear, with coincidence detectors. Activation at an ear passes down the delay line at a fixed rate beginning when the sound impacts the ear; in the illustration, this happens first for the right, then for the left ear. The coincidence detectors identify when the signals cross. The location of the sound is encoded by which coincidence detector is activated, as shown by the shaded gray circle. A neural circuit that implements this particular algorithm has been identified in the barn owl's brain stem.

Here we see how productive it can be when a system is studied simultaneously from multiple perspectives: computation, algorithm, and implementation. As is found in the barn owl, answers at each separate level can be linked together to form a coherent explanation of the whole system. You can see also, though, that such linkages are far from simple; differences between the field mouse and the owl show that answers at one level of description do not necessarily determine how the system operates at another level.

My third, and last, example is about language. Finally. Specifically, I want to give us a hint of how complex these linkings between computation, algorithm, and implementation can be in the case of language. Consider the simple sentence "Ada loves Winston." The gray box in panel A of Fig. 2 illustrates the grammatical structure of this sentence: It has a verb phrase ("VP") containing the verb "loves" and a noun "Winston"

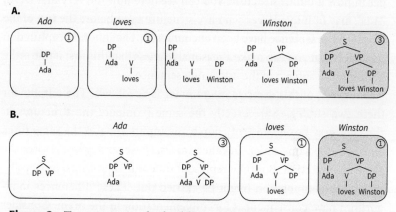

Figure 2. Two strategies for building structure. Two different algorithmic strategies are illustrated for how the brain might compute the grammatical structure for the sentence "Ada loves Winston." (A) illustrates a non-predictive "bottom-up" strategy that only builds structure after all words in the sentence are encountered. Circled numbers measure the amount of grammatical structure that is "built" at each step. (B) illustrates a more eager "top-down" strategy that builds structure in anticipation of upcoming words. Boxes delineate the syntactic structure that is built after each word in the sentence is encountered. Note that the final grammatical structure, shaded gray, is identical for both strategies.

(marked "DP" – don't worry if this abbreviation is unfamiliar to you). The verb phrase is combined with the subject of the sentence, "Ada". The question is this: When you encounter such a sentence one word at a time, how does your brain build this structure? Let me be clear, this is a *big* question which will occupy three chapters later in this book. But for now let's just consider a small piece of the puzzle. One strategy for how to build this structure is illustrated in the rest of panel A. Here, the structure is built after all the words that belong to the sentence have been encountered. You can see this by examining the individual boxes from left to right; the lines indicating the presence of a verb phrase (VP) or full sentence (S) are created only after the last word of the sentence is encountered. But this is not the only possible strategy for building this structure. Indeed, do you typically wait until someone has finished speaking before starting to make sense of what they are saying? I suspect not! Panel B of Fig. 2 illustrates an alternative strategy which is more eager in how it builds structure. You can see here how the very first word, "Ada", may fit into a larger sentence structure even before the rest of the words in the sentence have been encountered. The intuition captured in panel B is that you can make a reasonable guess that this first name is the grammatical subject of the upcoming sentence.

The point of this exercise in grammar is as follows. The outputs for these two strategies are exactly the same (compare the structures in the light gray boxes) and, of course, the inputs are also the same three words in the same order. In other words, these strategies perform the same computation. But they represent different algorithms, or steps by which the computation might be carried out. These differences in algorithm have consequences for implementation in the brain. Consider a simple (and common) implementation hypothesis that there is more brain activity (say, more neurons firing) when the brain is building more syntactic structure. As you can see, you can't test this hypothesis without also making a commitment about whether structure is built late in the sentence (as in panel A) or early (as in panel B). Moreover, the implementation hypothesis will also depend on just how syntactic structure works; in fact there are debates between linguists concerning, for example, the proper structure of a verb phrase in English.[5]

There are a few lessons to learn from these three examples. One takeaway is that research in neurolinguistics requires deep thinking about language at all of these levels: the computational problem being solved, the algorithm that might solve it, and the implementation of that algorithm. Another takeaway is that answers at one level of description do not necessarily depend on answers at another. David Marr (1982, p. 28) puts it this way:

> [F]inding algorithms by which [a computational theory] may be implemented is a completely different endeavor from formulating the theory itself. In our terms, it is a study at a different level, and both tasks have to be done.

A third key takeaway is that answers at these separate levels can and must be brought together into a more comprehensive account of a full cognitive system. Of course, bringing these together requires not just hypotheses at each separate level but also hypotheses about how the levels connect to each other. How, for example, does a particular algorithm for echolocation relate to banks of neurons? Or, how does the brain's implementation of a particular grammatical structure influence amount of brain activity when a certain word is encountered? These are *linking hypotheses*.

Neurolinguistics is, at its heart, a discipline concerned with these linking hypotheses. It's reasonble to say that a good bit of the rest of this book is an expansion on this one point. To make that point, you'll also be introduced to the tools of neurolinguistics and to a whole host of state-of-the-art results that have emerged from their use.[6] It's my hope that this will offer a foundation for you to discover, read, and engage with more of this research yourself.

With these aims we've set ourselves a tall challenge, but we aren't starting from scratch. The search for linking hypotheses between language and the brain builds on a tradition of research that goes back over 150 years.

A brief history

Nina Dronkers, a scientist at the University of California at Berkeley, offers a wonderful take on the very beginnings of the cognitive neuroscience of language. In a 2007 paper she and her colleagues revisit two case studies that were first presented by the neurologist Pierre Paul Broca in 1861 to the Anatomical Society of Paris.[7] In brief, the story goes like this. Broca was called on to examine a patient named Leborgne who had suffered a severe stroke many years earlier. As a result of the stroke the patient could no longer speak and retained only the ability to utter single syllables. Leborgne died shortly after this examination; this afforded a unique scientific opportunity, as Broca was able to study the damage to Leborgne's brain very close in time to the careful observation of Leborgne's linguistic behavior. At that time, autopsies were the only scientific tool available to directly study the human brain. The examination revealed damage that appeared to affect just one area of Leborgne's brain: a portion of the frontal lobe of the left hemisphere of the cortex.

Shortly after, Broca examined a second patient with a similarly severe deficit in producing speech. And an autopsy of the second patient revealed a startlingly similar pattern of brain damage. The same area of the left hemisphere's frontal lobe appeared to be damaged in both patients with severe difficulty in speaking. The story deserves a much longer telling, as the autopsied brains were set aside for preservation, lost when part of a building collapsed, found, lost, and found again in the cellars of the Paris School of Medicine in 1979.[8] As a result, these brains have now been subject to careful study by all the tools made available to modern neuroscientists. A picture of Leborgne's brain can be seen in Fig. 3A. The image shows clearly the frontal lesion in what is now often called *Broca's area* in honor of those revolutionary observations.

Why were Broca's observations revolutionary? A major debate in the medical sciences of the 19th century was whether, and in what way, the brain might be divided into parts that perform different functions. The *localizationist* view held that, yes, different parts of the brain support qualitatively different functions. This particular viewpoint had been most strongly argued by an earlier neurologist named Franz Josef Gall

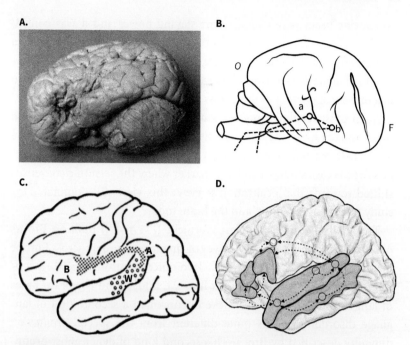

Figure 3. Over 150 years of language in the brain. (A) A case study published in 1861 provided Paul Broca with evidence that specific areas of the brain, such as the left frontal lobe, are connected to specific cognitive capacities, such as speech. (B) In the 1870s Carl Wernicke and students hypothesized that language relies on crucial links between the frontal and temporal lobes. (C) The neurologist Norman Geschwind popularized this "classical model" of how frontal and temporal areas of the brain work together to support language comprehension and production. (D) This diagram adapted from a 2012 review paper shows just how much modern accounts maintain the foundations of the classical model, with important novel insights.

Sources: A: Dronkers et al. (2007, fig. 3); B: Wernicke (1874, fig. 3); C: Geschwind (1970, p. 941); D: Adapted from Friederici (2012, fig. 1).

who, in the early 1800s, had published influential work that linked specific behaviors and personalities with specific brain regions. Gall's theory was based on measurements of the shapes and bumps of the skull. That particular way of measuring brains, called *phrenology*, is nonsense.[9,10] Still, the assumption that specific functions were *localized* to specific brain regions had some staying power and it was considered a viable, yet unproven, idea in 1861. Unproven, that is, until Broca presented to the Anatomical Society of Paris those two case studies showing that language itself could be severely disrupted following damage to just one small portion of the brain's left frontal lobe.

As in most retellings, my version simplifies things considerably and many other physicians and scientists contributed to that great debate over whether brain functions might or might not be localized. Nonetheless Broca's case studies stand as a marker where the scientific consensus shifted towards the localizationist view; this view still dominates the study of human cognition and the brain today.[11]

The Viennese physician Carl Wernicke (['vɛɹnəki]) built on Broca's observations by conducting a survey of language disorders across Europe in the latter decades of the 1800s. Two outcomes from this work continue to have a major impact on how we understand the brain bases of language today. First, Wernicke documented a second kind of language disorder that was quite different from the speech *production* difficulty described by Broca. This second kind of deficit most prominently affected speech *comprehension*. Whereas Broca's patients could comprehend instructions and answer simple questions, Wernicke studied brain-damaged patients who had great difficulty making sense of what was being said to them. In further contrast, these patients faced no obvious difficulty producing speech; they produced words and sentences with apparent ease and great fluency. But that speech was not, on the whole, sensible. That is, while fluent, the speech these patients produced seemed to be lack any coherent meaning. Like Broca, Wernicke used autopsies to link this behavioral difficulty with language comprehension and meaning to a particular pattern of brain damage; such patients typically suffered brain damage to the rear of the temporal lobe of the left hemisphere. This region, sometimes called *Wernicke's area*, is marked "W" on the diagram in Fig. 3C.

The second impactful outcome is that Wernicke's survey led to what has come to be called the "classical model" of language in the brain. One of Wernicke's diagrams for this model is shown in Fig. 3B. (Inexplicably, the diagram shows the brain's right hemisphere, but the model most certainly focuses on the left hemisphere for language.) I want to highlight three important pieces of this account. The first is a basic division of labor between two "language centers": an area in the temporal lobe that is necessary for comprehending speech and an area in the frontal lobe that is necessary for producing speech (marked A and B, respectively, in Fig. 3B). The second point is that these regions are necessarily connected. The language we produce is (absent neurological damage) also a language we understand. Third, these centers are closely connected to other brain areas whose functioning is also crucial for full use of language; these include motor areas for articulating words, memory areas for storing conceptual knowledge, and others. Even as scientists were first localizing distinct language-related brain areas, they also recognized that such areas are not isolated; their functioning depends crucially on deep interconnections with each other and with other areas of the brain.

Language disorders of the type documented by Broca and Wernicke that are due to brain damage are called *aphasias*. Table 2 summarizes the two such aphasias discussed thus far. There are many other fascinating syndromes that will be discussed in later chapters, as well as important complications and nuances to this kind of research (if you just can't wait, then skip ahead to page 27 in Chapter 2). Still, for over 100 years our understanding of the brain bases of language relied almost exclusively on this sort of research. The classical model first articulated by Wernicke was popularized and expanded upon by the neurologist Norman Geschwind (1970; 1972) in the middle of the 20th century.

Table 2. **Aphasia and the classical model.** Aphasias, or language deficits caused by brain damage, formed the key source of evidence for the "classical model" of language in the brain.

	Behavioral symptoms	Typical site of brain damage
Non-fluent ("Broca's") Aphasia	Slow, laborious, non-fluent speech • Short utterances • Comprehension relatively intact	Left inferior frontal gyrus ("Broca's Area")
Fluent ("Wernicke's") Aphasia	Fluent well-articulated speech • Familiar intonation patterns • Word meanings are disordered or inappropriate	Left posterior middle temporal gyrus ("Wernicke's Area")

Drawing on many decades of aphasia research, Geschwind argues that Wernicke's insights about localization largely stand the test of time, and moreover receive further support from patterns of aphasia documented since the 19th century that follow when the connections between relevant language regions are damaged. One such connection is the *arcuate fasciculus*, which is a bundle of neural fibers that connects the posterior temporal and inferior frontal lobes (see Fig. 3C). (By the way, please don't worry if some of these anatomical terms seem opaque to you. Getting oriented to the brain's "geography" is one of the main goals for the next chapter.)

The tools of modern neuroscience that are discussed in the next chapter go far beyond what was available to the pioneering neurologists of the 19th century, or even the scientists of the mid-20th. It is no surprise, then, that our understanding of the brain bases of language has changed significantly as scientists have confronted different kinds and ever-greater amounts of data not only on language disorders, but on the dynamic changes of brain activity that can now be recorded in real time while people use language. But it is difficult to overstate just how important that "classical model" is as a foundation for making sense of even the most modern accounts of language in the brain. One such moden example is given in panel D of Fig. 1. This illustration reflects one influential perspective developed by the cognitive neuroscientist Angela Friederici (2012). There are important components and nuances in this diagram that are absent from earlier models (panels B and C), but I hope you can also see that the underlying structure is largely the same: The brain bases of language are understood to involve key areas of the left hemisphere of the brain, specifically regions within, and connections between, the frontal lobe and the temporal lobe which are shaded in gray.

There is one way, though, that modern neurolinguistic models, which will occupy our attention for the rest of this book, differ quite radically from that "classical model." If you think back over the last few paragraphs you may well notice that "language" was discussed from two perspectives: that of language *production*, which was associated with *Broca's aphasia*, and that of language *comprehension*, linked to *Wernicke's aphasia*. There is a fundamental division, on this view, based on the whether the brain is making sense of some input or producing some sort of output. But since the time of Broca and Wernicke there has been a revolution in linguistics and the other sciences of the mind that centers on the mental stuff that stands between inputs and outputs: the stuff of cognition itself.[12] This cognitive view recognizes that the ability to use language reflects a kind of knowledge; it is what you know when you "know a language." Moreover, language is not just one kind of knowledge, but rather is made up of a whole range of mental entries, from knowledge of how to articulate the sounds or signs of a particular language, to the meanings and connotations of words, to the often implicit rules of grammar, and more.

The cognitive perspective asks us to look at how the brain represents and uses these different kinds of knowledge. We've come, you might have noticed, full circle in this chapter. Distinguishing the different kinds of knowledge that make up language from how that knowledge is represented and used by the brain lines up neatly with the distinct levels of description described in Table 1.[13] Indeed, even the very preliminary data we've seen so far resonates with this view. The language disorder documented by Carl Wernicke is, fundamentally, a disruption of meaning; patients show impaired language comprehension, but also the utterances they produce, despite being quite fluent, lack coherent meaning. The fundamental issue is not one of comprehension or production, inputs or outputs, but rather concerns something about that aspect of language that allows the brain to represent and make use of meaning.

The rest of this book

The rest of the book is concerned with trying to connect insights from linguistics regarding "what language is" with insights from cognitive neuroscience concerning "how the brain uses language." This linking between different levels of a complex cognitive system is exactly what makes neurolinguistics an especially challenging area of study and, in my view at least, a very rewarding area as well. Here's how we'll (try to) tackle it.

Chapter 2 offers a quick overview of the tools and methodologies from neuroscience that will be so crucial on our journey. Be warned that this chapter is heavy on the terminology, but "cheat sheets" are offered in the form of diagrams and, especially, Table 3 on page 41. Chapters 3 and 4 focus on sound: How does the brain create meaning out of sound waves in the air? How the brain accesses and represents those meanings is the focus of Chapters 5 and 6. We don't (typically) communicate with just single words. Language is remarkable, and remarkably unique, in how words are combined together to create say and think things that, quite literally, no one has ever said before. Chapters 7–9 turn to this area. The ideas that we discuss across these chapters are inter-related. While I have written them with a particular ordering in mind, I have also tried to sprinkle breadcrumbs liberally so that connected ideas are cross-referenced with each other; in this way those of you who want to skip around can do so without too much trouble.

Chapter summary

This chapter has laid out the groundwork for what this book is about. We were introduced to a particular way of thinking about brain systems in terms of three different levels of description, the linkages between each level, and a little bit of the history for how scientists have studied "language in the brain."

- To ask good questions of a cognitive system like language, we must distinguish between different levels of description: the *computational* goals of a system, the *algorithmic* steps needed to meet those goals, and the *implementation* in a physical system to carry out those steps.
- *Linking hypotheses* capture how possible answers at each of these levels connect to each other.
- Efforts to specify these links go back over 150 years, starting with research on *aphasia*, or language disorders caused by brain damage.
- *Non-fluent aphasia* describes a difficulty producing fluent speech associated with damage to the brain's left frontal lobe. *Fluent*

aphasia is a difficulty with understanding and producing sensible language; it is associated with damage to the brain's left temporal lobe.
- Early aphasia research led to the *classical model* for language in the left frontal and temporal lobes which still influences modern theories.

To tackle the challenges laid out in this introduction, we need to get a handle on the tools and techniques of neurolinguistics. These are the focus of the next chapter.

2

The toolbox

To get started we'll need to become familiar with the terminology and tools used to investigate brain function in people. The goal here is to get just enough technical details under our belt that we can make sense of why a researcher might choose to use a certain tool to answer a question they have in mind. There is obviously a lot of fascinating nuance that we won't have time for here – notes sprinkled through the chapter are there to point you towards resources to help you dig deeper into the wonderful world of brain-measuring tools.

Brain geography

The first thing to discuss is the brain's anatomical "geography." Like stepping off the train in a new city, it pays off to take a moment to orient yourself before striking out. Scientists approach anatomical orientation from three points of view; you can think of these as if you were looking at a brain from three different directions. The *sagittal* view presents the brain as seen from the side; the *axial* view presents the brain as seen from the bottom, and the *coronal* view presents the brain as seen from the back. Each of these views are shown in Fig. 4.[1]

Now, in each view we can talk about whether something is higher or lower, towards the front or back, etc. On the vertical dimension, higher areas are *superior*, also called *dorsal*, while lower areas are *inferior* or *ventral*. Areas towards the front are *anterior* while areas towards the back of the brain are *posterior* (less frequently, you may also see anterior and posterior areas described as *rostral* or *caudal* respectively). Finally, one can draw attention towards the left and right sides of the brain, or *lateral*

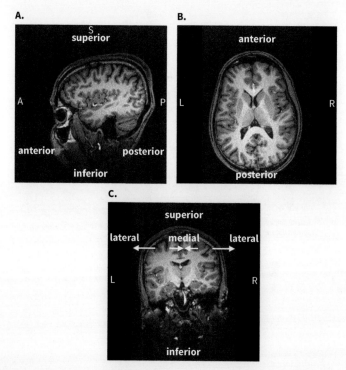

Figure 4. Three views of the author's brain. (A) sagittal, (B) axial, and (C) coronal.

areas, in contrast to *medial* (or "mesial") areas towards the middle. All of these orientation terms are illustrated in Fig. 4.

With these terms in hand, we can turn our attention to some major anatomical features and landmarks of the human brain. The central nervous system is divided into three components: the *brain-stem*, the *cerebellum*, and the *cerebrum*. The latter is the site of much of the complex perceptual, cognitive, and motor-related processing done by the brain and it's where we'll focus our attention in this overview. The cerebrum is divided into a left and right hemisphere, and comprises an outer shell, or *cortex*, which surrounds a number of sub-cortical structures. A sagittal image of the medial aspect of the human cortex is shown in Fig. 5A.

Perhaps most striking on first looking at the human cortex is just how *convoluted*, or folded, it is (Fig. 5B). The peaks of these convolutions are called *gyri* (singular: "gyrus") while the valleys – these are the parts

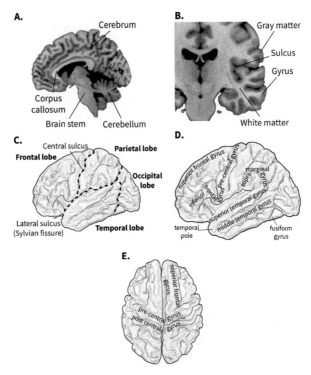

Figure 5. The central nervous system. (A) Major parts of the central nervous system; the *corpus callosum* is a bundle of fibers that connects the left and right cortical hemispheres. (B) Coronal cross-section of the cortex highlights its convolutions and the distinction between gray and white matter. (C) The major lobes of the cortex. (D–E) Some major macro-anatomical features of the cortex.

that you can't see without spreading the folds apart – are called *sulci* (singular: "sulcus"). These gyri and sulci help to define some of the principal large-scale, or *macro-anatomical*, landmarks of the human cortex.

The first such anatomical landmarks are the four *lobes* of the cortex. About mid-way between the anterior and posterior poles of the cortex is a deep sulcus that starts at the top of the brain and extends downward and laterally. This is the *central sulcus* and it serves to divide the *frontal lobe*, which is anterior to the sulcus, from the *parietal lobe*, which is posterior. Make sure you can find each of these features in Fig. 5C.

At its bottom, the central sulcus intersects with another major dividing element: the *lateral sulcus*, which is also called the *sylvian fissure*. This sulcus runs perpendicular to the central sulcus, and extends anteriorly. The lateral sulcus divides the more superior frontal lobe (which you've already found) from the *temporal lobe*, which lies inferior to both the frontal lobe and the parietal lobe. (Notice how we are practicing with the orientation terms, like "superior" and "inferior.") Lastly, at the posterior edge of the cortex lies the *occipital lobe*. Unlike the other three, no major sulci divide this lobe from either the temporal lobe, which is more anterior, or the parietal lobe, which is more superior. You can see the approximate divisions between these regions in Fig. 5C.[2]

The two gyri on either side of the central sulcus merit special mention. On the anterior side is the primary *motor cortex*, while on the posterior side is the primary *somatosensory cortex*. These two gyri are complementary: The motor cortex sends signals to the peripheral nervous system to control muscles, and the somatosensory cortex receives signals related to touch. The neurons within these gyri are organized into two maps of the body. So, for example, neuronal populations controlling the feet are clustered in a location that is separate from neurons controlling the hands, neck, or face. In the motor cortex, the tongue, being unique both in its flexibility and its sensitivity, is separately represented from other neurons controlling other aspects of the face and vocal tract. This spatial separation between regions involved in different areas of the body and, indeed, vocal tract will prove a source of insight in later chapters.

Still with me? The last few paragraphs have been very heavy on terminology, and I'm afraid we are not done yet. The figures are provided to keep you grounded: Try to use the terminology so far to label a few arbitrary points on Figs 4 and 5. For example, can you put your finger on the anterior edge of the frontal lobe of the cortex? How about a medial-superior point on the right parietal lobe? Take a moment to check your understanding before moving on.

In fact, the common labels for many cortical landmarks are based on almost exactly the scheme you've just practiced with. For example the "superior temporal gyrus" simply refers to the most superior gyrus on the temporal lobe of the cortex. There are two other gyri on the temporal lobe, which are sensibly called the "middle temporal gyrus" and "inferior temporal gyrus." The same logic applies to sulci (can you point to the "superior temporal sulcus"?) and it also works for the frontal lobe. So, the "inferior frontal gyrus" is the most inferior of the three gyri on the frontal lobe (try to trace this gyrus with your finger on Fig. 5 moving from anterior to posterior).

But, matters aren't quite this simple. First, researchers typically abbreviate anatomical regions, so "superior temporal gyrus" is shortened just to "STG"; "IFG" by the same logic means "inferior frontal gyrus." Abbreviations may include more information as well, so "PITG" or "pITG" (you see both conventions) means "posterior inferior temporal gyrus" and so on. Second, there are many anatomical regions that retain less conventional, more idiosyncratic names. For example, the "supramarginal gyrus" ("SMG") spans the meeting-point of the temporal, frontal, and parietal lobes and curves along the edge (or margin) of the lateral sulcus. Just posterior to the SMG in the parietal lobe is the "angular gyrus" ("AG"), named for its shape. A third example is the gyrus that lies on the inferior side across the temporal and occipital lobes: It is called the "fusiform gyrus" – again, a reference to the shape of this particular structure.

I recommend three things for students facing the great variety of terms introduced in this section: (1) Memorize a few of the locations that are most common in the particular domain of study (you'll see just what these might be in the following chapters for speech, words, and sentence-processing). (2) Get comfortable using the more general and "compositional" terms like "anterior superior frontal gyrus"; you can use these even when a piece of anatomy may have a different name. For example, say you forget the name "supramarginal gyrus", try "gyrus that curves around the posterior end of the sylvian fissure" instead. (3) Lastly, the internet is your friend: look up unfamiliar terms (I do!)

All of the terminology so far captures the brain's *macro*-anatomy – those features that you could see with your own eyes. Of course, what really makes the brain, well, a brain is the *micro*-scopic neuron. There are about 100 *million* of these basic building blocks in the human central nervous system. Each of these cells has a cell body, or *soma*, an *axon* which, like a long arm, reaches out (sometimes quite far) to connect with other neurons, and a large set of *dendrites*, which are smaller appendages

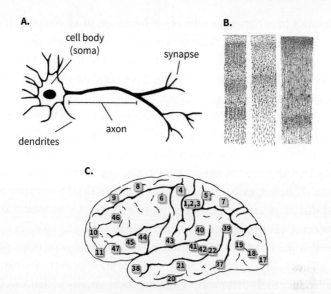

Figure 6. The neuron. (A) The parts of the neuron. (B) Neurons are organized into cytoarchitectural layers in the cortex. (C) Cytoarchitecture can be used to define regions, as in Brodmann's atlas.

Sources: A: Image modified from original by Looxix via Wikimedia Commons under the CC-BY-SA 3.0 license (https://creativecommons.org/licenses/by-nc-sa/3.0/); B,C: Public domain, via Wikimedia Commons

on the receiving end of some other neuron's axon. The meeting place of one neuron's axon and another neuron's dendrite is a small gap called the *synapse*. An illustration of a neuron is given in Fig. 6A.[3]

There are a few features of neurons that will be important for our purposes, including the electrical discharges that serve to pass information from one neuron to another, and the way that whole groups of neurons are organized into layers in the cortex. But there are many fundamental aspects of neurons that we will, generally, not be addressing here, such as neurotransmitters and the biochemical interactions at the synapse, neuronal cell types, and more. Why don't we discuss such fundamentals of the nervous system? Well, frankly, our current understanding of how the brain carries out language is just too coarse-grained to get much out of that level of detail. Current research doesn't yet have a whole lot to say about how these fine-grained neuronal properties relate to how we represent and use linguistic knowledge.

Neurons pass signals to each other by means of an electrical current, an *action potential*, that flows along the "sending" neuron's axon. This discharge evokes a chemical interaction at the synapse that in turn changes the electrical potential at the "receiving" neuron's dendrite. The effects of that discharge are summed together across the many dendrites of a particular neuron. Certain connections may be excitatory, which increases the probability that the receiving neuron will itself fire off an action potential. Or connections might be inhibitory and lower the probability that the neuron fires. Microscopic electrodes that measure the electrical discharges at individual neurons show that these cycles of electrical discharge and rest occur anywhere from once to over 250 times per second. (For reference, the microchip in the computer I'm using to write this paragraph carries out about 2.3 billion calculations per second.) These electrical interactions will become important shortly, when we introduce tools that measure the electrical activity of neurons during language use.

A second property of neurons that merits discussion is their organization in the cortex. Neuronal cell bodies are clustered at the surface of the cortex; this is the "gray matter" that gives the brain its characteristic hue. Underneath this sheet is the "white matter" – tissue made up of axons that connects neurons together: The whiteish color comes from *myelin*, which is an electrically insulating substance that coats many axons, thereby making electrical transmission more efficient. When put under a microscope, the cell bodies at the surface of the cortex are not uniformly or evenly distributed; rather, the cells are organized into six distinct layers. The layers differ from one another in terms of the principal cell types and their spatial arrangement. A drawing of these layers, from the pioneering 19th-century neurologist Santiago Ramón y Cajal, is shown in Fig. 6B. The systematic organization of neurons into cortical layers is called *cytoarchitectonics*.

Neuroscientists studying the cytoarchitecture of the brain have mapped a remarkable pattern of organization that complements the macro-anatomical features of sulci and gyri that were introduced above. These maps group areas of cortex together when they have similar neuronal organization. The leading idea here is that populations of neurons with a similar organization may be involved in similar functions; so such a map might provide interesting clues as to what areas of the brain form coherent regions that carry out a common function. One such map is shown in Fig. 6C. This comes from work by Korbinian Brodmann that was published in 1909 based on meticulous study of a single individual. For better or worse, this particular mapping, which divides the cortex into 52 distinct areas, has become commonly used by cognitive neuroscientists. So, just as a researcher might refer to the "aSTG" (see above), based on macro-anatomy, they may also refer to "Brodmann area 38", or just "BA38." Likewise, a researcher may refer to "BA45" or "BA4" which span, respectively, a small portion of the inferior frontal gyrus (IFG) and the gyrus immediately anterior to the central sulcus.

Summary so far

We have come to the end of the brief (and dense) overview of brain anatomy. Remember, your goal is to become comfortable navigating around images like the those shown in Figs 4 and 5. As part of this, I encourage you also to make use of the resources that were mentioned in the notes to the pages above.

Now that we know a bit of our way around the brain, we are ready to introduce the main tools that the neurolinguist uses to study how the brain carries out language.

Imaging brain structure with MRI

You may already be familiar with one of the principal technologies used to take images of the brain: *Magnetic Resonance Imaging*, or MRI. This same technology is used to make high-resolution images of broken bones and to map the location of tumors. When used to image brain tissue, MRI creates 3D images with high *spatial resolution*. And, as we'll see in a moment, MRI is a very flexible technology, allowing researchers not only to take images of brain structure but also to see how different regions connect to each other and even to measure how much oxygen different areas are using up as they function. What makes MRI so flexible, in part, is that it takes advantage of something that your body has plenty of: water.

Each water molecule in the tissues of your brain has two hydrogen atoms. Under normal conditions, the nuclei of those hydrogen atoms are oriented more or less randomly. MRI uses a series of very strong magnets, first, to align the nuclei of those atoms along a common axis and then to perturb those nuclei, knocking them out of alignment. This perturbation introduces energy into the system, and so as the nuclei relax back to their (aligned) resting state, that energy is released. This release of energy, or "resonance", is measured in two dimensions by tracking how this energy interferes with yet another magnetic field. A three-dimensional stack of these 2D images is then made by moving the magnetic field slightly and repeating the process.

The spatial unit of analysis in these images is the *voxel*, or "volumetric pixel": brain images are composed out of these basic elements, which typically measure 1–3 mm. To get a sense of scale here, note that a voxel of 1 mm^3 contains about 50,000 neurons. The distribution of hydrogen across bodily tissues allows this technique to distinguish, for example, the cortical gray matter which houses neuronal cell bodies, from white matter, which is made up of the axons that convey electrical signals from one neuron to another. Indeed, if you go back to Fig. 4, you can see clearly the contrast between the outer gray tissue and the whiter tissue it surrounds.

A wonderful example of the flexiblility of MRI comes from something called *diffusion tensor imaging*, or DTI. DTI is used to measure the structural connectivity between different regions of the cortex. These connections are instantiated by axons that project from one region of the brain to another; large groups of axons form bundles that seem like highways that connect distant cities. As you can see for yourself – look again at the white matter in Fig. 4 – standard MRI does not have the resolution to reveal the directions traveled by bundles of axons. But there is an ingenious solution based on MRI's sensitivity to the distribution of water. By orienting the magnetic fields in different ways, MRI can measure how well water diffuses in different directions. Well, water is more likely to diffuse along an axon rather than to cross cell boundaries between axons. Thus, by following the direction that water is diffusing, researchers can estimate the direction that axons are traveling.

By analogy, cars drive along roads, not between them. You could therefore reasonbly estimate a map of roadways by tracking the movements of cars.

Another example of the wonderful flexiblity of MRI is how it can be tuned to measure properties of blood flow. Before we get to that, I'd like to discuss how MRI brings a new perspective to the study of language deficits and brain lesions that was introduced in Chapter 1. These approaches, and other tools, are techniques that probe brain function, and this is where we now turn our attention.

Of deficits and lesions

The deficit/lesion method relies on a clinical evaluation to determine how someone's language might be impaired, and connects those deficits with the location of brain damage, or lesion. We've already seen examples of this method in action with the work of Broca and Wernicke discussed in Chapter 1. As in those historical examples, modern clinical practice identifies *aphasia* (remember: a language disorder due to brain damage) using a clinical evaluation. This evaluation might, for example, uncover a pattern of symptoms consistent with *non-fluent aphasia* (a.k.a "Broca's aphasia") such that patient produces only slow, short utterances. The diagnosis is always based on behavioral symptoms alone.

Unlike in Broca's day, scientists no longer have to wait for an autopsy to uncover what kinds of brain damage might be correlated with those symptoms. MRI is used to take an image of the damaged tissues *in vivo*; just such an image is shown in Fig. 7A where damaged neural tissue appears darker than surrounding gray and white matter (can you describe the location of the lesion using the anatomical terminology from earlier in this chapter?)

There are a number of different kinds of aphasias and related deficits. Rather than list them here to be memorized, I will be discussing specific examples in the chapters that follow.

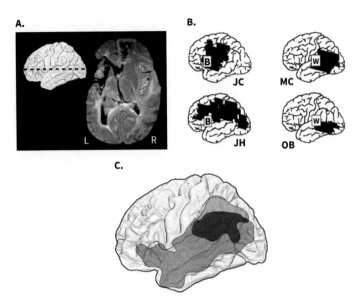

Figure 7. The deficit/lesion method. (A) MRI of Leborgne's brain; notice how the frontal lobe lesion, shown on the top left, penetrates quite far beneath the cortical surface. (B) On the left are two patients with non-fluent aphasia; only the top one has a lesion in the traditional "Broca's area"; the two right-hand patients have Wernicke's aphasia; B and W indicate traditional Broca's and Wernicke's areas respectively. (C) Illustration of overlapping lesions from two Wernicke's aphasia patients; medium and light gray shows non-overlap; dark indicates overlap.
Sources: A: Dronkers et al. (2007); B: Dronkers et al. (1999); C: Adapted from Rogalsky et al. (2011).

One reason deficit/lesion research is so valuable is that it can be used to build a *causal* argument. What this means is that a researcher can infer, under certain circumstances, that a particular brain region (say, the inferior frontal gyrus) is necessary for the brain to carry out a particular function (say, fluent speech). This kind of inference contrasts with *correlational* reasoning, which we'll see examples of below. The strongest way to build such a causal argument relies on what is called a *double dissociation*; here's what that means:

	Patient A	Patient B
Capacity X	Impaired	Not impaired
Capacity Y	Not impaired	Impaired

Let's consider a concrete example.[4] Patient VER had a stroke at the age of 68 that reduced blood-flow to frontal and parietal lobes in her left hemisphere. This stroke led to severe language deficits affecting both comprehension and production. Despite these limitations, she remained able to follow simple instructions. She was able, for example, to point to a picture that matched a spoken word ("point to the car"). Interestingly, she could match certain kinds of words more easily than others: she was particular good at matching words for kinds of food, but worse at words for common house hold objects. Now consider patient SBY, who was diagnosed with Herpes Simplex Encephalitus at the age of 48. This degenerative disease damages brain tissue in the temporal lobes. SBY retained fluent speech but showed difficulty with speech comprehension. In a picture-naming task, SBY showed difficulty with kinds of food but, fascinatingly, was much better at identifying pictures of household objects. (By the way, a neural deficit that affects one's ability to recognize things like food or household objects is a kind of *agnosia*. We'll discuss these sorts of deficits in more depth in Chapter 6.)

Here is a diagram comparing these two case studies:

	VER	SBY
Object words	Impaired	Not impaired
Food words	Not impaired	Impaired

This is a double dissociation! A pattern like this allows you to infer that at least some brain systems necessary for reasoning about household objects are distinct from those that are necessary for reasoning about food concepts (and vice versa). To understand the value of this, consider the *single dissociation* present in the case of patient VER alone. Having experienced serious neural trauma, it is possible that the patient has difficulty with more complex tasks or concepts, compared to simpler tasks or concepts – indeed, there are lots of reasons why objects might be more difficult sorts of concepts than food; perhaps household objects are used in more varied ways (for writing, cutting, cooking, stapling, switching...) than foods (eating). A double dissociation allows more precise reasoning. Both patients have experienced trauma, so the difference in performance must be due to the different areas affected by their brain damage.

You may already be thinking of some of the challenges facing the deficit/lesion method. For example, the damage experienced by the patients I've described is relatively broad. It spans the frontal, parietal, and/or temporal lobes (though, is it that unreasonable for food, of all things, to take up a large part of our brains?) Relatedly, the nature of the damage is different between each patient; in this example, one patient had a stroke while the other a neuro-degenerative disease. Even if two patients have the same source of damage (say, stroke), the nature of the neural trauma and the presentation of language difficulties would never be exactly identical between any two different people. When can two or more individuals be treated as "comparable"? Yet another challenge is that this method relies on brain damage that is subject to chance and to the idiosyncrasies of our (fragile!) bodies: Stroke damage predominantly falls along certain vasculature in the brain, and neuro-degenerative diseases target certain parts of the central nervous system. What this means is that there is no guarantee that the patterns of damage most often seen in the clinic will neatly isolate relevant parts of language processing.

These challenges are partially illustrated in Fig. 7B. Here, two patients with a clinical diagnosis of non-fluent ("Broca's") aphasia are shown on the left, and two patients with fluent ("Wernicke's") aphasia are shown on the right. First, notice the broad extent of the lesions. You can also see how different these lesions are from each other, even for patients with similar clinical diagnoses. Indeed, the patients in the bottom row do not have any damage to the brain regions that were historically linked with these two deficits under the classical model discussed in Chapter 1 (see Fig. 3 on page 11).

In some ways, these limits are addressed by complementing deficit/lesion-based research with the other tools in the toolbox that will be discussed below. But there are also some strategies for analyzing lesion data that can help with at least a few of these downsides. One such strategy is the *lesion overlap* method which is illustrated in Fig. 7C. The lesion overlap method takes advantage of MRI's spatial resolution to tackle the challenge that patients differ both in the location of brain damage and in their clinical symtoms. For example, researchers may identify a group of patients who, like patient SBY, have difficulty with food-related words. Now, these patients may also differ in other respects (e.g. the precise location of their lesion, whether they have other co-morbid deficits in language, memory, etc.). Keeping these differences in mind, researchers can use high-resolution MRIs of each patient to identify the location of each lesion, and then look for which specific areas overlap between all of the individuals who, despite differing in many ways, share an impairment with food-related words. If researchers find a region where damage is shared between all individuals who show difficulty with food words, and that region is also preserved in control participants who don't have difficulty with food words (but have other aphasia symptoms), then such a result provides compelling support for linking food concepts (in this example) with a specific sub-part of the brain which generalizes across a group of patients.

Blood flow and function

There are a growing number of tools that let scientists measure brain function *in vivo* – while those brains are using language in some way. I'll first talk about *functional imaging*, which means taking pictures of where brain regions are activated, usually by tracking patterns of blood flow. Then, I'll talk about *electrophysiology*, which measures the rapid changes in electrical currents that are generated by neurons.

By far the most widely used tool in all of neurolinguistics is *functional MRI* (fMRI). Remember, MRI is super-flexible, and that flexibility extends to measuring properties of blood flow. Here's how that works: When the brain is at rest there is a certain balance of oxygen in the brain's vasculature – in your bloodstream. When neurons become more active they demand more oxygen, removing it from the bloodstream.[5] Technically, the balance of oxygenated hemoglobin (oxygen-carrying

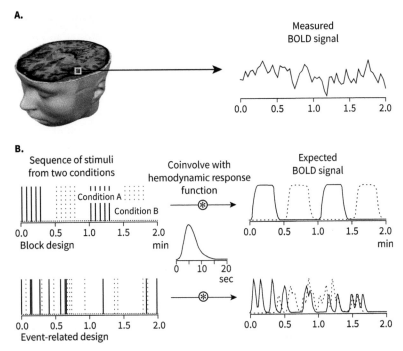

Figure 8. The fMRI method. (A) Example of the fMRI BOLD response from one voxel. (B) A schematic of stimuli for *block* and *event-related* experimental designs (left); the stimulus events are coinvolved with the hemodynamic response function (middle) which yields an expected BOLD signal for each condition (right).

blood cells) goes down and de-oxygenated hemoglobin goes up. Your metabolic system responds to this dip by flooding the blood vessels surrounding the active neurons with more oxygen (in the form of oxygenated hemoglobin). All together, the dips and peaks in *blood oxygenation* follow the pattern shown in the middle of Fig. 8B. This is the *hemodynamic response function* and it is not very speedy! Indeed, the influx of oxygenated blood can take over six seconds to respond after brain activity has begun.

So, why this business about neural metabolics? MRI, by virtue of measuring properties of hydrogen atoms, is sensitive to changes in blood oxygenation. This dynamic change in blood oxygenation creates an MRI signal called the *blood oxygenation-level dependent*, or BOLD, signal, illustrated in Fig. 8A. You'll recall that MRI has good spatial resolution, and that's true for the BOLD signal as well. So, fMRI offers a way to measure, within just a few millimeters, the location of brain activity. It does so at the cost of being slow; the hemodynamic response unfolds over several seconds; fMRI has low *temporal resolution*. This sluggishness is quite different from the millisecond speed of neural activity itself, and also slower than the speed of language; you might hear anywhere from two to six words a second in everyday speech.[6] A second limitation of fMRI is that it is quite loud, and this is especially important for speech studies. But there is a clever solution that comes from embracing these two limitations. The idea is that the participant listens to a speech stimulus in relative silence, and the MRI is only turned on afterwards. This works because the BOLD signal associated with processing the speech stimulus peaks several seconds after the stimulus has begun.[7]

Despite this sluggishness, the BOLD signal changes in proportion to the amount of underlying neural activity (Boynton et al., 2012). This means you can still tap into neural activity associated with speedier processes, like language, provided you do so thoughtfully. This relationship is schematized in Fig. 8B for two different experimental setups. One setup is a *block design* experiment. Here, the individual stimuli from each condition are presented all together, one after another. The right side of Fig. 8B takes into account the sluggish BOLD response – this is simply what you would expect the fMRI signal to look like if it precisely followed your experimental design. The researcher then statistically tests how well the measured BOLD signal from a particular voxel (8A, right) matches up with your BOLD signal predictions (8B, right).

As you can see on the top row of Fig. 8B, a block design makes it easy to tell apart the BOLD signals associated with each condition, despite the fact that hemodynamic changes are pretty slow. The bottom of Fig. 8B shows an *event-related design*. Here, the stimuli from each condition are interleaved. Still, when they are carefully spaced, the BOLD signals corresponding to the two conditions can be teased apart (Fig. 8B, right-hand side, bottom).

There are a number of other tools that take images of brain function using properties of blood flow. Together, these are called *hemodynamic* techniques. These include *functional near-infrared spectroscopy* (fNIRS), which measures the BOLD signal – just like fMRI – but does so using light (it turns out that oxygenated and deoxygenated blood scatter light differently). Unlike fMRI, fNIRS just involves a small cap containing the light emitters and receivers; it is quiet and more comfortable than fMRI, making it especially appropriate for studies with children (Rossi et al., 2012). Another tool that you may see mentioned is *Positron Emission Tomography* (PET). PET used to be quite common but has mostly been supplanted by fMRI. Rather than tracking blood oxygenation, PET measures glucose uptake. Just like oxygen, neurons that are active demand more glucose from the blood stream. The glucose is tracked using a radioactive marker that is injected into patients before each block of experimental trials. Among the downsides to this approach is the fact that it takes over one minute for the most commonly used radioactive marker to decay, meaning PET is an even slower technique than fMRI – you must use a block design for PET experiments.

The electric brain

Another strategy for measuring brain function is to record the electrical activity of neurons themselves. Together, these are *electrophysiological* techniques. The most common strategies are non-invasive (no brain surgery) and involve using sensors placed on or near the scalp to measure electrical and magnetic fields that are generated by tens of thousands of neurons acting together. These techniques together have high *temporal resolution* because they measure neural activity as quickly as it occurs. But because they measure that activity from outside the scalp, they only have moderate to limited *spatial resolution* when it comes to determining where in the brain the activity comes from. In certain circumstances, doctors implant electrodes directly onto neural tissue (this is done, for instance, to help guide surgery to resolve certain kinds of epilepsy). These (thankfully) uncommon instances are incredibly valuable for researchers, as they pair high temporal resolution with very precise spatial resolution.

You may have heard of *electroencephalography*, or EEG. This is a very common electrophysiological tool. It works by using electrodes placed on the scalp. The current generated by a single neuron is so tiny as to be unmeasurable from outside the head. But, if neurons fire together – in synchrony – and they are oriented so that they all generate currents that face in the same direction, then those currents sum together, making a signal that *can* be measured by something as simple as a conducting metal placed onto the scalp.[8] More precisely, the signals measured by EEG (and also MEG, to be introduced below) reflect not the action potentials themselves, but post-synaptic electrical discharges from suitably aligned populations of neurons. It turns out that a great many of the neurons in your cortex are aligned in this way.

The raw EEG signal combines many different electrical signals, including the activity of cortical neurons that are doing something of interest to you (say, reading this paragraph), but also signals from other brain activity (like the part of your brain that's thinking about what you'll eat for lunch) and even the often much stronger signals generated by moving your neck muscles, blinking your eyes, the buzzing of overhead lights, or even the elevator in the room down the hall. The most common technique for separating out the EEG signal that is relevant for an experiment from all these sources of noise is to average together the EEG signal from many repetitions of the same kind of experimental event. This averaging technique is so common that it gets a special name: the *Event-Related Potential* (ERP).

Fig. 9A–B illustrates how an ERP is created from raw EEG data for an experiment with two conditions. When the data from each condition are averaged together relative to a common event – here, the onset of a new written word – certain systematic patterns that were hidden in the raw signal become apparent. These systematic voltage fluctuations, shown in Fig. 9B, are ERP *components*. Components have four properties:

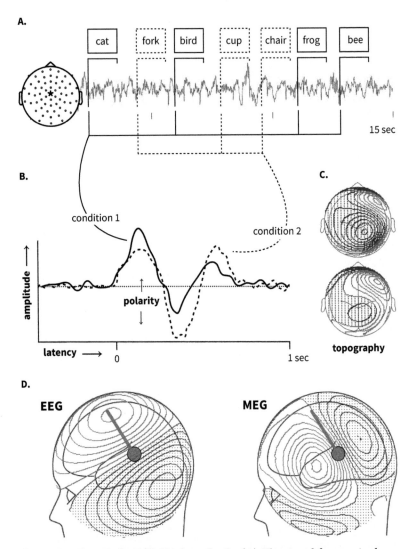

Figure 9. Electrophysiological methods. (A) The signal from a single recording channel can be divided into epochs based on stimulus events. (B) Averaging the data for events from the same condition shows yields the *Event-Related Potential*. This components of this potential can be described in terms of their *amplitude, latency, polarity,* and (C) their distribution in space, or *topography*. (D) Electroencephalography (EEG) and magnetoencephalography (MEG) measure complementary signals generated by neuronal currents; MEG signals are easier to localize in space because they are less distorted by the scalp.

- *amplitude* – how much the voltage changes (shown on the y-axis in the plot),
- *polarity* – the direction of change,
- *latency* – when the voltage changes (shown on the x-axis),
- and *topography* – where on the scalp the voltage was recorded.

I need to say a few things about how ERPs are shown graphically. First, there is an old tradition of plotting negative voltages up, and postive voltages down, on the *y*-axis of an ERP plot. But this tradition is not always followed. My advice: Read the axis labels carefully. Second, a line-plot usually shows an ERP from one single electrode (or maybe the average of a few neighboring sites). This view is good for illustrating amplitude, polarity, and latency, but is not so useful for topography. A topographical plot, shown in Fig. 9C, offers a view of an ERP topography at a particular point in time.[9]

While the ERP is the most common way to analyze EEG data, it is far from the only strategy. We'll come across a variety of other analysis techniques later in the book when we look at specific examples.[10] EEG is a very widely used technique because it is both relatively inexpensive (unlike, say, MRI) and it can also be used with a wide range of people, including even very young infants. However, beause the technique relies on sensors that are placed directly on the scalps of research participants, standard methods for EEG data collection can be more difficult from participants with thick and/or coarse hair (Etienne et al., 2020).

The electrical currents which make up the EEG signal also generate magnetic fields. *Magnetoencephalography* (MEG) measures these fields, though doing so is technically trickier (and more expensive) than EEG. Briefly, MEG involves a helmet filled with liquid helium to super-cool electrical coils. These coils pick up the tiny changes in magnetic flux generated by neuronal activity. Those changes in magnetic flux are truly tiny, being roughly 100,000 times smaller than the field generated by a car across the street, 1,000 times smaller than a heartbeat, and over 10 times smaller than the current generated by your cellphone when measured from the other side of the table. Why go to this trouble if MEG measures the same electrical activity as EEG? The primary reason is that magnetic fields, unlike electrical currents themselves, are easier to localize in space. This means that MEG has higher spatial resolution than EEG while keeping the same (high) temporal resolution.[11]

A comparison of the signals generated by MEG and EEG is shown in Fig. 9D. Here, the underlying neural current is shown by an thick line. The topography of voltage that is measured by EEG is shown on the left, while the topography magnetic flux measured by MEG is shown on the right. One thing you'll notice is that the EEG voltage is very spread out. This spreading occurs because the electrical current passes through many different tissues as it travels from the cortex out to the scalp. These tissues have different conductivities which distort and spread the signal. In contrast, magnetic flux travels through the tissues of the head with very little distortion. And, because the shape of a magnetic field is lawfully related to the current which generates it, the location of that current can be reconstructed, or *spatially localized*, with reasonable accuracy based on the topography of magnetic fields recorded outside the head.

Aside from spatial localization, many other aspects of the MEG signal are analyzed in a similar manner to EEG. For example, MEG signals are typically averaged together to remove noise sources, creating the *Event-Related Field* (ERF).

Under rare circumstances, medical patients are implanted with electrodes in a procedure called *electrocorticography*, or ECoG. This can occur, for example, when surgeons are monitoring for epileptic activity prior to surgery. Sometimes, this monitoring lasts for several days, and in these cases, researchers may gain a patient's consent to perform an experiment. Depending on medical necessity, electrodes are either placed on the surface of the cortex or may be inserted into deeper layers of the cortex or into sub-cortical structures (sometimes called *intracranial electrodes*, ICE, or stereoencephalography, sEEG). The data recorded from this procedure is valuable medically and scientifically. The electrical signals are recorded directly from adjacent neuronal populations (within 1 millimeter), meaning that ECoG has very high spatial *and* temporal resolution. Further, because only local electrical potentials are recorded, the data has a much higher ratio of signal to interfering noise than other electrophysiological techniques. Finally, the ECoG signal can record neural signals with much higher frequencies than those recorded by M/EEG (higher-frequency signals have much lower power than low-frequency signals; only signals below 80 Hz or so are strong enough to record outside of the head). As you can see, these sorts of measurements can be quite exciting to researchers. But keep in mind that ECoG shares with the deficit/lesion method the limitation of relying on a patient population that is not randomly selected, and whose neural activity is (by definition) atypical. These limitations should be kept in mind when seeking to draw generalizations from ECoG data.

Stimulating and inhibiting brain function

Let's talk about one last set of tools for probing brain function before wrapping up this tour of the neurolinguist's toolbox. Rather than passively recording brain activity, these tools alter brain activity, either by stimulating it or by inhibiting it. This allows researchers to examine how such effects change language processing.

One such tool is *Transcranial Magnetic Stimulation*, or TMS. Briefly, TMS uses a strong and focused magnetic field to induce or inhibit electrical currents in the cortex. TMS can be targeted fairly precisely, affecting a few millimeters of cortical tissue. Different applications of TMS yield different kinds of results. For example, if TMS is applied in a single pulse, this can induce an action potential (when applied over the hand part of the motor cortex, this pulse can make your finger move). When applied repeatedly (so-called "rTMS") at a relatively low frequency, say once per second, this pattern tends to inhibit neural activity in the targeted region, creating a so-called (temporary) "virtual lesion"; faster rates are thought to enhance neuronal excitability.[12]

Another stimulation tool is *Direct Cortical Stimulation* (DCS). DCS involves a current, not a magnetic field, and is applied directly to the cortex of a surgery patient. This is an invasive technique, and it is commonly used in tandem with ECoG (described above) as a way to monitor patients undergoing brain surgery. For example, patients will be asked to name pictures while the surgeon stimulates parts of the temporal lobe; the surgeon will then note sites that affected fluent speech. While invasive, DCS is incredibly powerful, as it allows the surgeon to infer with incredible spatial precision whether a part of the cortex is necessary for a certain behavior.[13] A non-invasive variant of this technique called *transcranial direct current stimulation* (tDCS) has also been developed with applications as a clinical tool in the treatment of aphasia or other neurological syndromes.[14]

The great advantage of stimulation techniques is that, like deficit/lesion methods, they allow inferences about whether one region is causally related to a particular language outcome. Because these techniques don't rely on particular patient populations, TMS is much more flexible and, potentially, can lead to more generalizable results. Stimulation tools also have great temporal resolution, as pulses can be applied at different time-points during an experimental trial. Still, as with deficit/lesion methods again, our limited understanding of neural plasticity is a barrier to more fully understanding the outcome of stimulation methods, especially those that interfere with processing.

Chapter summary

We've now opened the toolbox and seen what's available for probing the neural bases of language. We have learned the terms we need to navigate around the brain and to make sense of someone who says "I applied TMS to the anterior portion of the left inferior frontal gyrus."

Each of the techniques that we've reviewed has both advantages and disadvantages for making sense of brain structure and function. Becoming familiar with these trade-offs is perhaps the most important lesson from this chapter; it's these features that link up particular methods to specific research questions. For example, if you have a research question concerning the timing of linguistic events ("Are semantic or syntatic representations accessed first during comprehension?"), then you'd better use a technique with the right temporal resolution (quick: Which ones might work?). On the other hand, if you have a research question about whether linguistic stimuli might involve the same representation ("Are two allophones encoded as a single phoneme?") perhaps you might consider a technique with high enough spatial resolution to test for similar

Table 3. **Methods cheat-sheet.** Summary of the main pros and cons for each of the techniques for measuring brain function reviewed in this chapter

Technique	Summary	Pros	Cons
deficit/lesion	Links patterns of brain damage with language impairments	Earliest technique for mapping brain function • *Double dissociations* allows causal inferences	Relies on specific patient populations • Difficult to generalize; no two lesions or patient symptoms are alike • Brain plasticity/recovery poorly understood
fMRI	Measures blood oxygenation changes that correlate with neuronal activity	High spatial resolution • Non-invasive • Commonly available	Low temporal resolution • Loud
fNIRS	Measures blood oxygenation using scattered light	Moderate spatial resolution • Cheaper than fMRI • Quiet • Good for pediatric studies	Low temporal resolution • Lower spatial resolution than fMRI
PET	Measures changes in blood glucose	High spatial resolution • Developed prior to fMRI • Not subject to some magnetic interference that affects MRI	Very low temporal resolution • Involves radiation • Generally supplanted by fMRI
EEG	Scalp electrodes measure electrical currents from cortical neurons	High temporal resolution • Relatively inexpensive • Non-invasive • Quiet • Good for a variety of populations	Very low spatial resolution

Continued

Table 3. *Continued*

Technique	Summary	Pros	Cons
MEG	Measures magnetic fields generated by cortical neurons	High temporal resolution • Moderate spatial resolution • Non-invasive • Quiet	Expensive • Less common than EEG or fMRI
ECoG	Measures neuronal activity with electrodes implanted in the cortex	Exceptional spatial and temporal resolution • High signal-to-noise ratio	Highly invasive procedure • Only used when medically indicated • Recording site based on medical need • Relies on specific patient populations
TMS	Uses a magnetic field to excite or inhibit cortical activity	High temporal resolution • Moderate spatial resolution • Allows causal inferences	Brain plasticity poorly understood
DCS	Uses electrical current to directly stimulate neuronal activity	High temporal and spatial resolution • Allows causal inferences	Highly invasive procedure

or different activation patterns (such as...?). To help you, Table 3 aims to bring together the main pros and cons for each of the techniques we've discussed.

Of course, to really get to grips with all of these new concepts, we need to use them. So, we now turn to how this whole suite of tools has been used to understand how the brain comprehends speech.

3
Sounds in the brain

How does sound traveling through the air become brain activity associated with a word's meaning? We'll tackle this question in this chapter and the next in terms of three separate transformations. Each takes us one step forward on the pathway from sound to meaning:

acoustic signal →[1] neuro-auditory representation (neurogram) →[2] neuro-phonologcal representation (phonological sketch) →[3] lexical item

Spatial and temporal codes for sound

Before we get going, be sure not to confuse *speech* comprehension, the main topic of this chapter, with *language* comprehension. Speech is just one of several *modalities* that language may use. The study of sign language comprehension provides deep insight into neural principles of language that are shared across modalities.[1]

At the beginning of its journey, the spoken word is all but unrecognizable. Speech comes to the listener as vibrations in the air impacting your eardrum. Those vibrations have already been shaped by the outer ear and ear canal to emphasize some frequencies over others. On the other side of the eardrum is a viscous fluid that bathes a most remarkable feat of evolutionary engineering: the *cochlea*. Coiled like a snail, the cochlea contains the perceptual organ responsible for converting sound waves into electrical impulses for the brain (see Fig. 10). This conversion is carried out by thousands of tiny hair cells. They move when the surrounding fluid vibrates, triggering a cascade of neuro-chemical processes that sends action potentials along to the central nervous system.

Language and the Brain. Jonathan R. Brennan, Oxford University Press.
© Jonathan R. Brennan (2022). DOI: 10.1093/oso/9780198814757.003.0003

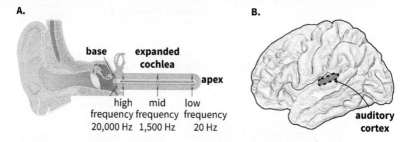

Figure 10. Auditory pathway. (A) Schematic of the human ear and cochlea; different sites along this structure respond to different sound frequencies. (B) The location of the primary auditory cortex.

Source: (A) Adapted from an image by OpenStax, via Wikimedia Commons, under the CC BY 4.0 International license (https://creativecommons.org/licenses/by/4.0/).

The cochlea is exquisitely structured as it tapers from a large base towards a narrow apex. This shape means that different frequencies are emphasized at different points along the spiraled organ: high frequencies at the base, and low frequencies at the apex. Thus, hair cells at different locations of the cochlea are moved by different frequencies. This is a *spatial code* – different information is represented by different neurons. Spatial coding is the first of several principles underlyng how the brain represents sound.

If you were to plot out the action potentials generated by hair cells moving at different locations of the cochlea, you would see something that is remarkably familiar from Introductory Linguistics: a spectrogram showing changes in sound energy at different frequencies over time (a spectrogram is shown at the bottom of Fig. 11). The shape of the cochlea's spectrogram is specially tuned to the kinds of sounds humans have evolved to hear, from about 20 Hz, or twenty vibrations per second, up to 20,000 Hz (depending on your age and propensity for loud music). We can call the neural responses that represent sound at the brain's periphery the *cochleagram*.

Brain signals generated at the cochlea pass through a series of subcortical structures before arriving at a bit of cortical tissue that projects up from the superior temporal gyrus. This is *Heschl's gyrus* or the primary auditory cortex (sometimes abbreviated "A1"). Like the visual system in mammals, the auditory pathway mostly goes to the opposite,

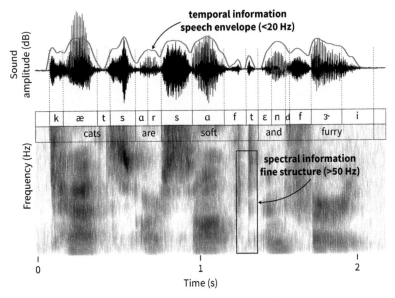

Figure 11. Speech information unfolds on multiple time-scales. The speech waveform (top) shows relatively slow changes in loudness following the speech intensity envelope (gray line) that reflects syllabic information. Spectral information (bottom) carries fine structure detail that distinguishes phonemic features, like the rapid change from [f] to [t] (black box).

Source: Modeled after Giraud and Poeppel (2012b, fig. 9.1)

or *contra-lateral*, hemisphere of the brain; sounds going in your left ear are mostly processed by the auditory cortex in your right hemisphere. All together, it takes about 50 milliseconds for electrical signals from the cochlea to register in the auditory cortex.

How is sound represented in the auditory cortex? The first thing to look for is whether the spatial code created at the cochlea is preserved in the auditory cortex. Indeed, studies in animals and in humans have shown that different parts of the auditory cortex respond selectively to different frequencies. This cortical spatial code for sound is called *tonotopy*. In one elegant study, researchers played simple tones as they were swept from low to high and high to low frequencies over 60 seconds (Talavage et al., 2004). Using fMRI, they monitored for peaks in the BOLD signal in the auditory cortex. These peaks traveled as the frequency swept from high to low, revealing multiple regions within the auditory cortex that show a gradient "map" for tonotopy.

The connection between a certain stimulus, like a tone at 1000 Hz, and the response of a given neuron is called the *receptive field* of that neuron. So, another way to talk about tonotopy is to say that neurons in the primary auditory cortex have receptive fields that are tuned to different frequencies, and neurons with similar tuning are adjacent to each other. This notion of a receptive field has proven to be very fruitful in other domains, like vision, to understand how neurons process information.[2]

Before we continue with the intricacies of the brain's auditory system, we need to turn our attention to the speech signal itself. How, indeed, could we make sense of the brain bases of speech comprehension if we don't consider just what makes something, well, speech? Familiar concepts from phonetics, such as phonemes and distinctive features, will be introduced shortly. First, let's consider just how remarkably flexible humans are when it comes to recognizing speech. It turns out that frequency information – our main focus so far – is not always that important for understanding. Speech scientists break down an acoustic signal into two parts, shown in Fig. 11. This familiar waveform has peaks and troughs – points in time where the sound is louder or softer. Tracing these peaks and troughs is the *speech envelope*; it's what gives speech its rhythmic patterns of syllables and stress contours. Zooming in shows the rapid sound vibrations that make up the speech energy at different frequencies; this *fine structure* includes narrow bands of energy, like the formants that distinguish vowels, but also broad-band noise associated with sibilants and fricatives. Quite interestingly, speech comprehension is actually fairly good even when there is very little fine structure, and only the speech envelope.

One way to test for the relative importance of fine structure is to use a technique called *noise vocoding*. In vocoding, the speech envelope is extracted from an utterance, and the fine structure is replaced with white noise (equal sound energy at all frequencies). When this procedure is done in a very coarse way, say using the envelope from just ten bands ranging from the lowest to highest frequency (a range from 20 to over 10,000 Hz), the speech is still remarkably understandable.[3]

So, the brain needs not only to represent frequency information, the *spectral* signature of speech, but also to capture *temporal* information, or the changes in loudness over time.

There is some really neat evidence that the brain tracks temporal information – the speech envelope – in exquisite detail. In one representative study, researchers used ECoG recordings from the left temporal lobe, including auditory cortex, while participants simply listened to natural speech utterances (Kubanek et al., 2013). When they examined the signal coming specifically from electrodes in the auditory cortex, they found a remarkable match between the rapid fluctuations of brain activity and the equally rapid fluctuations of the temporal envelope. This match between brain activity and perceptual input is called *entrainment*, and may be a key method the brain uses to "lock on" to important elements of the sensory system for further processing.

The word "entrainment" almost implies a kind of mind control in that the brain automatically locks on to percetual inputs. But envelope entrainment, in fact, is subject to control by your attention. This is pretty important, because the speech that we *comprehend* is only a fraction of the babble that we *hear*. I'm writing this paragraph right now from an airport lounge where I can hear 3–4 conversations, two different radios, and a whole range of other noises like fans, motors, etc. How do we follow speech amidst all this noise? A familiar example of our resilience to complex noisy environments is the *cocktail party effect*: This is the experience of being in a crowded room and noticing when we hear our name spoken somewhere on the other side of the room. It's as if our name just "leaps out" of the babble of the busy room.

There are at least two ways we could navigate the chaos of real-life acoustic inputs: We might "hear" all of it at early stages, including the overlapping acoustic input from multiple sources, and then rely on higher-level linguistic processing to sort out which words to attend to. Or, alternatively, even early perceptual processing might be more selective, such that we only hear just that part of the acoustic input that we consider important. The data are consistent with the second of these two views. One piece of supporting data comes from a study by Nai Ding and Jonathan Simon at the University of Maryland 2013.

They had participants listen to segments of speech from an audiobook that were mixed in with different amounts of noise during MEG recording. The noise was crafted to have a similar spectrum to the speech stimulus – kind of like people talking over each other in a busy room. At the loudest levels, the noise made it so the participants could not understand the audiobook at all. At moderate levels the noise still strongly distorted the acoustic input itself, but participants could understand the speech robustly. Moreover, the researchers could use patterns of auditory cortex neural activity to reconstruct the "true" speech envelope, even when noise distorted that envelope in the stimulus that participants heard. So, temporal processing in the auditory cortex can be selectively tuned to separate out specific speech sounds from other acoustic inputs. It's as if we can turn up the volume on just the most relevant parts of the bubbling noise in our environment.

How might the temporal information from the speech envelope be encoded in the auditory cortex? Brian Barton, Alyssa Brewer, and colleagues (2012) devised an fMRI experiment that built on the use of frequency-swept tones for tonotopy, already mentioned above. The experiment had two kinds of acoustic stimuli (see Fig. 12B). First were simple bursts of noise that were restricted to a range of small frequency bands (350–450 Hz, 750–850 Hz, 1550–1650 Hz, etc.; think *very* artificial formants); this tested for tonotopy much like the study above. Second were stimuli with white noise that changes in loudness periodically (it is *amplitude modulated*); the sounds get louder and softer at a specific rhythm, like two, four, or eight times a second. You'll remember from Chapter 2 that fMRI is loud (see page 33). To help the participants hear these stimuli, the researchers used a technique called *sparse sampling*: What that means is that the fMRI is only turned on after the stimuli have already been presented. This works because the BOLD signal representing the brain's response to those stimuli takes six or more seconds to reach its maximum (for a reminder, take a look back at Fig. 8 on page 32).

Using these stimuli, the experimenters first mapped how different parts of the auditory cortex responded to sounds at different frequencies. As expected, this showed the now-familiar tonotopic pattern; an example of their tonotopy results is shown on the top panel of Fig. 12C.

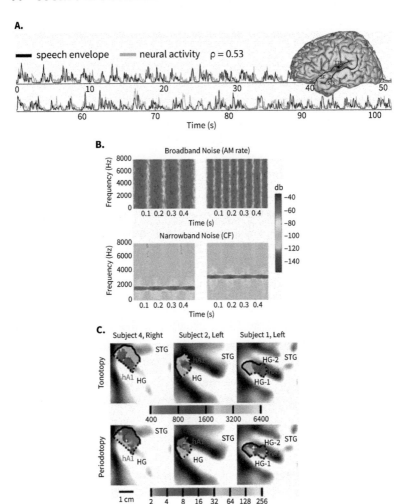

Figure 12. The neurogram. (A) Neurons in the primary auditory cortex track speech envelope information. (B) Stimuli used to map temporal (top) and spectral (bottom) spatial codes in the primary auditory cortex. (C) Primary auditory cortex from three subjects shows gradient spatial maps for spectral (top) and temporal (bottom) information.
Sources: A: Kubanek et al. (2013); B and C: Barton et al. (2012).

They then mapped out the auditory cortex response to the periodic stimuli. Just as with tonotopy, different parts of the auditory cortex respond to different kinds of periodic stimuli, and it does so in a gradient: Neurons that respond to a 2 Hz stimuli are adjacent to those that respond at 4 Hz, which are next to the 8 Hz neurons, etc. This is shown in the bottom panel of 12C. You can call this new spatial code *periodotopy*. Moreover, the spatial arrangment of this response was remarkable: The neurons that responded to changes in periodicity were almost exactly 90 degrees offset from the neurons that responded to changes in frequency. It looks as if the auditory cortex contains a tiny little x- and y-axis (actually, several sets of axes) that encode both the spectral and temporal details of incoming stimuli.

We've begun to get a picture of something we can call a *neurogram*. This is the "brain's-eye-view" of sound. Sound is characterized by spectral patterns – sound energy at different frequencies – and also by temporal patterns of loudness varying over time. These patterns are encoded with a fine degree of detail in the auditory cortex. We've now reached the first stop on the pathway from sound to meaning.

But (there's always a "but"), just because the brain can represent the acoustic details of speech doesn't yet tell us how it can discover the words that are carried by those sounds. Words are built up of *phonemes* – speech sounds. And phonemes themselves are built up from *distinctive features*. The striking thing about phonemes and features is that they are *categorical*: A phoneme is either a /t/ or a /d/ ("mat" or "mad"); there is no in-between! But, what we've seen so far about the neurogram is not categorical, it is continuous: Neurons respond to changes in spectral or temporal information in gradient patterns, from low to high or fast to slow. How does the brain turn the continuous acoustic information encoded in the neurogram into discrete speech information?

Neurograms and the phonological sketch

To find our way from the continuous world of acoustics to the categorical world of speech and language, we need a clearer grasp of the puzzle facing the brain. First, we will see concrete evidence that the neurogram alone isn't sufficient for understanding words. Then we'll see that the early auditory responses in the brain indeed do represent categorical information, right alongside continuous information. Once we are firmly puzzled by these fascinating findings, we'll then turn to some clues for how the brain carries out the remarkable transformation from neurograms to phonemes.

I'll start with two classic pieces of evidence that speech perception can't be understood in terms of continuous information alone. Both pieces of evidence have to do with the great variation in speech signals and how listeners seem to handle this variation with ease. The first piece of evidence is, straightforwardly enough, the phenomenon of *categorical perception*. Briefly: You start with speech sounds that have been artificially altered, say starting with the syllable [sa] and changing the acoustics incrementally until you reach a [ʃa] syllable. While such stimuli form a smooth continuum, listeners don't perceive a smooth change. Rather, they typically report hearing a sequence of [sa] sounds followed by a sequence of [ʃa]s.[4]

Now, you might think that all the /s/ tokens that you hear – despite their differences from each other – form a nice neat acoustic cluster that is different from related sounds (/ʃ/, /z/...). But that is definitively not so; this is one of the real striking puzzles confronting the brain's ability to understand speech. For example, if listeners hear a word with a sound artificially altered so as to be amibiguous between, say, [s] and [ʃ], they will report hearing whichever sound makes a sensible word. So, [wɪ_] is typically heard as "wish", but [lɛ_] is heard as "less" (the underline "_" indicates the acoustically altered in-between sound). This phenomenon is called the *Ganong effect* (Ganong, 1980). A similar effect is found when listening to different speakers. If listeners are cued to *think* they are listening to an adult male speaker of American English, they are more likely to hear [s] sounds for words that begin with an ambiguous [s]-or-[ʃ] sound ("said" vs. "shed"), and the reverse is true when a listener thinks the speaker is female.[5] Somehow, listeners are able to navigate the great variability in speech with apparent ease – even to the point that one speaker's [s] is another's [ʃ]!

These pieces of evidence indicate that continuous acoustic information – what we saw encoded in the brain's neurogram in Fig. 12 above – isn't enough to explain how the brain recognizes phonemes and, ultimately, words. Corroborating insights come from brain data. One particularly fascinating example is an ECoG study by Nima Mesgarani, Edward Chang, and colleagues (2014) at the University of California, San Francisco. In this study, patients with electrodes implanted in their superior temporal lobe – including the primary auditory cortex – listened to sentences that were selected to include a wide range of English phonemes. By gathering signals from different sites along the cortex, the researchers identified locations that responded specifically to certain phonemes. The first thing they found was that, indeed, electrodes at different sites were selective to specific phonemes. So, for example, one electrode only responded to stop consonants ([d] [g] [b] [p] [k]...), another to sibilant fricatives ([s] [z] [ʃ]), and another to nasals ([n] [m] [ŋ]). You can see some examples of their data in Fig. 15C on page 65. These sort of data indicate that populations of neurons in the superior temporal gyrus have *phonemic receptive fields*.

This is cool, and it gets cooler. First, these brain responses were *fast* – starting just 100 milliseconds after the beginning of a phoneme (remember, it takes about 50 milliseconds for auditory information just to get to the cortex). Second, these brain responses weren't organized in terms of phonemes themselves, but actually in terms of the distinctive features that phonemes are built from. We already see this in the examples above: An electrode might be sensitive to the class of stop consonants, not just to specific phones like [d] or [k].

So, early neural responses in the temporal lobe reflect phonological features. Are these responses really categorical? "Hold it right there!" says the skeptical reader. "Stops and fricatives have very different acoustics. *Of course* their neurograms won't be the same, and because neurograms are a spatial code we'd expect ECoG electrodes at different locations to respond to different phonemes."

The best evidence that these brain responses are indeed categorical comes from electrodes that show the classical non-linear response associated with categorical perception. While individual consonants vary in the acoustics of voicing, Mesgarani and colleages see a specific electrode that responds in an "all or nothing" way when the consonant is voiced, another that is "all or nothing" when the consonant is unvoiced, and yet a third that responds to stops regardless of whether they are voiced or not.

These data show how even early auditory responses capture categorical, not continuous, features. In addition, other data show that high-fidelity continuous acoustic information in the auditory cortex isn't enough to guarantee speech comprehension. Nourski and Brugge (2011) measured how well the brain tracks speech envelope information using ECoG data gathered from patients with electrodes planted directly into their auditory cortex. Their measurements show that the speech envelope is accurately tracked in the auditory cortex, even as the stimuli were sped up to about three times their original rate. Recall from page 47 that speech perception itself remains fairly high even when participants mostly only recieve envelope information. Despite the fidelity to the speech signal found in the auditory cortex, the participants themselves began to have great difficulty understanding the speech when it was sped up by a factor of about 2.5 (you can try this yourself next time you listen to an audiobook). So, despite the auditory cortex following even very rapid speech envelope changes, this doesn't mean that listeners can perceive that speech.

The data that we've seen over the last few pages deepens the speech perception puzzle facing the brain in a few specific ways. Speech perception is categorical, not continuous (see categorical perception, Ganong effect, categorical responses from temporal lobe electrodes). These categories can't fully be understood as summaries of acoustic differences (see the Ganong effect) and also, even when the auditory cortex can capture acoustic details, this doesn't guarantee speech comprehension (see sped-up speech). Our guiding question is then: *How* does the brain transform a continuous neurogram to a categorical representation of speech sounds?

Let's take stock of the ingredients we have for finding an answer: (i) the input we have to work with is the neurogram, (ii) the outputs we need are phonemic features, (iii) the transformation occurs relatively relatively rapidly (by 100–200 ms after speech begins), and (iv) it is carried out in the superior temporal lobe, in the vicinity of the primary auditory cortex. The next ingedient comes from one of my very favorite studies.

Listeners can make sense out of a great deal of speech variability, but we're stumped when speech is played *backwards* – it sounds totally alien! Even backwards speech, though, can become comprehensible in the right circumstances, and these circumstances put us on the path towards one answer for the puzzle at hand. In a paper just five paragraphs long, Kourosh Saberi and David Perrott (1999) describe a study in which participants listened to reversed speech. The stimuli were modified so that the reversals spanned larger or smaller windows of time. To do this, the sound waveforms were divided into segments of, say 300, 200, 100, 50, or 20 milliseconds each. Then, each segment was reversed, and the segments were put back together in the original order. In all cases the participants heard reversed speech. But the segments, or "grains" of speech, that were being reversed were either large or small. Listeners couldn't make sense of the words when the reversals spanned large segments of 300 or 200 milliseconds. When segments were smaller than 200 ms, participants started to make some sense of what they heard. Remarkably, the listeners reached 100% accuracy when the reversed segments were still about 30–50 milliseconds long. To put this result in perspective, the difference between voiced and voiceless stops – [ta] versus [da] – hinges on acoustic differences that last just 20–30 milliseconds. So, reversed speech is almost perfectly comprehensible if the grains of sound that are being reversed are small enough, where "enough" is about 1/20th of a second.

These data point to the leading that the brain has a *sampling rate* for speech sounds. What that means is that the brain takes a "snapshot" of acoustic information at fixed windows; everything that unfolds during that snapshot is integrated into one chunk of information; and that information is used to pick out phonemes. Researchers have proposed a few variants of this general idea, which I'll call *temporal windows of integration*.[6] We'll focus on two such theorized windows. The first spans roughly 25–40 milliseconds, which is about the window size identified in the speech reversal experiment above. This window is about the right length to distinguish the fine structure spectral properties of different phonemes. A second window is longer, spanning 200–300 milliseconds. This latter window is about the right length to sample the properties of syllables and other structures that make up the temporal envelope of speech.

Is there evidence that the auditory cortex can sample, or integrate, auditory information in these time-scales? Much of the relevant data comes from the study of *neural oscillations* – these are brain waves, or periodic patterns of brain activity that occur at specific rhythms (you've probably heard of one of these, alpha waves, which are associated with a relaxed awake state). Brain waves emerge when groups of neurons synchronize their activity such that they become excited and/or inhibited together. Such synchronization is important for processing information; it aligns whole populations together in terms of when they optimally receive input and send output on for further processing.

In a 2007 study, Anne-Lise Giraud and colleagues measured neural oscillations at different rates using EEG, and combined these data with spatially precise fMRI data.[7] They focused on *endogenous* oscillations – these are the brain wave patterns present when you are at rest; they aren't associated with any particular external (*exogenous*) stimulus. From the EEG data, the researchers extracted changes in signal power in a band of frequencies, including from 28–40 Hz (a 40 Hz signal completes one cycle every 25 milliseconds) and also from 3–6 Hz (166–333 milliseconds per cycle). Then they examine fMRI data collected simultaneously for changes in the BOLD signal that correspond to changes in these oscillations. They find that increased oscillations in just this band of frequencies correspond to changes in the BOLD signal right in the auditory cortex. And the auditory cortex response is *asymmetrical*: The left hemisphere shows a relatively stronger response to the shorter "phonemic-feature-sized" oscillations, and the right hemisphere shows a stronger response to the longer "syllable-sized" oscillations.

The data suggest that the auditory cortex has intrinsic rhythms, and these rhythms are matched in time to linguistic features of speech on two levels. The shorter window is appropriate for integrating spectral information that is used to identify phonemic features. The longer window appears to be well-suited to identify temporal properties of carried by syllables, including stress and intonation patterns.

If these two rhythms are crucially involved, in some way, in sampling the continuous acoustic input, then the asymmetry we just saw should mean that listeners prefer different kinds of speech information in each hemisphere: Auditory input to the left hemisphere should show higher performance – better speech comprehension – when fine structure is preserved, and the right hemisphere should show higher performance when the speech envelope is preserved. Houda Saoud, Anne-Lise Giraud, and colleagues (2012) test just this prediction using an experimental method called *dichotic listening*. In this sort of experiment, isolating earphones are used to present different auditory stimuli to the left and right ears (recall that the *left* ear ⤳ *right* hemisphere, and *right* ear ⤳ *left* hemisphere). With this setup, researchers present spoken words that are artificially manipulated to remove either the speech envelope or fine structure detail for either ear. First, the study reports some now-familiar facts: Words that have been modified to reduce their fine structure detail are still relatively well understood, but words that are modified to remove their envelope are almost impossible to understand (see page 47). When fine structure and envelope information is presented to *different* ears, comprehension is improved but, crucially, comprehension is best when there is a match between the underlying oscillations and the kind of information being presented at each ear. That is, listeners show better performance when fine structure information is going to the left hemisphere (faster "phonemic-feature-length" oscillations), and envelope information going to the right hemisphere (dominated by slower "syllable-length" oscillations).

We've skipped over a crucial piece of data in our discussion so far. What is the evidence that the brain integrates information in that larger 200–300 millisecond window? I've connected this time-window to syllables and to the temporal envelope of speech. But what is the evidence that the brain takes samples – snapshots – at this larger time-window? A striking piece of evidence comes from the *McGurk effect*. (I suspect this will be at least a little familiar to many readers.) We've mostly been focused on the processing of speech alone. The McGurk effect arises when auditory and visual information is combined together. Classically, listeners watch a video of someone while they speak a single syllable; the stimulus is edited so that the audio and visual signals don't match, for instance, by combining audio of someone saying [ba] with a video of that same person saying the syllable [ga]. Confronted with this sort of stimuli, listeners typically report hearing [ga] or, sometimes [da]. It is very rare for a listener to "hear" a [ba] – even though it is indeed a [ba] sound that was presented (McGurk and MacDonald, 1976).

Virginie van Wassenhove and colleagues (2007) examine what happens when the auditory and visual information aren't presented at exactly the same time. They present mismatched acoustic-visual stimuli with different time-delays, or lags, between the two stimulus parts. For example, the acoustic stimuli might be presented 200 or 100 milliseconds before the visual, or 100, 200, 300... after the beginning of the visual stimulus. They then record which stimuli were "McGurked" by the presence of mismatching visual information – that is, when did participants "hear" something other than the actual auditory stimulus? When plotted out as a function of time, van Wassenhove et al. observe that the McGurk effect is quite robust to mismatches in time; participants experienced the illusion even when the auditory stimuli occurred up to 200 or so milliseconds after the beginning of the visual stimulus! This pattern of data suggests that the brain does indeed integrate speech information that is spread over 200 milliseconds apart; here, audio and visual information is integrated into a single syllable percept.

The puzzle the brain faces is how to map continuous input to categorical phonological information. The answer we're considering is that it does so by taking samples, or snapshots, of the input at different temporal windows – integrating information at time-scales tuned to phonological and syllabic information. There are many questions we can ask about the this account. One chief question is how the system handles variable and degraded input. Indeed, we already saw on page 52 that speech perception seems to involve more than the straight mapping of acoustic to phonological information from the "bottom up". How does non-acoustic information facilitate speech perception?

There is a good deal of information about speech that is not carried in the acoustic signal we've been discussing so far. The visual information causing the McGurk effect, discussed just above, is one such source. Another source of information are the expectations of the listener. Listeners make predictions about what a speaker might say next; just think about all the times you've jumped in and finished someone else's sentence. You take advantage of many kinds of information when you do this, such as what you already know about your language (possible words, possible speech sounds), what you know about the speaker herself (what she might be likely to say), and what you know about the immediate linguistic context (what you are talking about).

Another way to put this idea is that listeners have the knowledge to internally *synthesize* upcoming speech. Generating such an internal representation of what they expect to hear can help in the subsequent *analysis* of the speech signal when it comes. Such an account of speech perception is called, appropriately enough, *analysis by synthesis*. A good example of how the brain uses predictions for speech perception is found in a study by Ediz Sohoglu, Matthew Davis, and colleagues at the University of Cambridge in 2012. They presented participants with single words that were distorted to a greater or lesser degree (they used noise-vocoded stimuli, which we already encountered above on page 47). This distortion made it moderately to extremely difficult to identify phonemes from acoustic information alone. Participants also saw a written word briefly flashed on the screen prior to the auditory stimulus; the word either matched or didn't match what they heard.

A matched written word presented before a distorted spoken word helped participants accurately identify what they heard. Did this improvement actually affect the mapping from acoustic to phonological information itself? (It's possible that the written word helped participants make a decision about what they heard without actually affecting the perceptual process.) The evidence suggests that predictions help refine phonological processing; when the word and audio matched, there was a very early brain response in the left frontal lobe, within the inferior frontal gyrus, and this brain response was followed by reduced activation in the superior temporal gyrus. It's as if predictions that were generated in the frontal lobe eased the burden on the temporal lobe when it was confronted with a distorted acoustic input.

We see here that existing linguistic knowledge plays an important – perhaps central – role in speech perception. The way this knowledge is combined with the input to help with recognizing words is illustrated in Fig. 13. This observation connects to a much larger issue of whether understanding speech requires a kind of perception that is different from how we perceive other sensory stimuli. I think evidence from

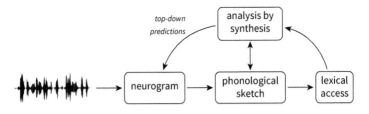

Figure 13. Analysis by synthesis. Acoustic information is mapped to neural representations of sound – the neurogram – and then to phonological representations. This *analysis* is aided by *synthesis* of speech input based on top-down knowledge.

neurolinguistics offers new insight into the old and contentious debate as to whether "speech is special." We'll circle back to this debate in the next chapter.

Chapter Summary

Take a breath. We've come a long way in this chapter – from sound waves moving hair cells in the cochlea all the way to phonological analysis.

- Acoustic information is represented with a *spatial code* both at the cochlea, and in the *primary auditory cortex*. That means that different neurons encode different acoustic features of speech. *Tonotopy* is the spatial code for frequency information, where neurons that are adjacent to each other respond to sounds with similar frequencies. The brain may also represent temporal acoustic information in a similar way, called *periodotopy*.
- The brain maps from these continuous neural representations of sound, or *neurograms*, to categorical linguistic units like phonemes to create a *phonological sketch* within about 100–150 milliseconds in the *superior temporal gyrus* surrounding the auditory cortex.
- This mapping is carried out by integrating acoustic information within at least two temporal windows: A shorter window for the

fine structure of speech that captures spectral detail, and a longer window that captures changes across time, such as the *speech envelope*.
- The phonological sketch is refined by a series of feedback loops between acoustic input and linguistic knowledge called *analysis by synthesis*. The outcome of these interlocking processes is a representation suitable for identifying words.

We've completed the transformations that map from acoustic input to brain representation of sound, and then to phonemes. The next chapter turns to the neural representation of phonemes themselves.

4
A neural code for speech

This chapter zooms in on the neural representation of phonemes. But, we have one thing to discuss before we get started. Consider: What is the "point" of speech perception? My answer may sound like a bit of a cheat, but it turns out that from the "brain's-eye-view" the point really depends on what the listener has been asked to do – their *task*.

My main focus for the rest of the chapter and, indeed, book will be just on one kind of task: finding meaning. On this view, the point of speech perception is to identify words, phrases, sentences, and so forth, in order to understand the speaker's message. "What else could you do?" you might ask. Well, rather than comprehending, you might be repeating, or shadowing, what you've just heard (think about relaying a phone conversation to someone standing next to you). Another task comes up in a laboratory settings, where a researcher might ask you to listen to input and press a button if you hear a certain phoneme (this is a *phoneme monitoring task*).

I mention these alternatives because they have a large effect on what happens to speech information once it's passed through the early stages of perception that we've seen so far. Fig. 14 sketches a *dual-stream model* of this larger speech-processing system developed by Gregory Hickok and David Poeppel (2007). On this view, there are two pathways. One is for understanding; this *ventral stream* travels along the temporal lobe and then the frontal lobe, carrying out transformations from acoustics through to phonemes, then lexical items, and ultimately the meaning of sentences. The ventral stream is shown in dark gray in Fig. 14. If you are engaged in a non-understanding task like speech repetition, however, then this theory suggests an alternative *dorsal stream* for processing.

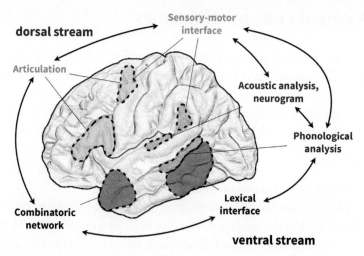

Figure 14. A dual-stream model for speech perception. Acoustic information in the primary auditory cortex interacts with phonological representations in the posterior superior temporal gyrus. Then, speech input is hypothesized to travel along two paths depending on the listener's task. For tasks involving comprehension, processing moves anterior along the temporal lobe and to the inferior frontal lobe; this is the *ventral stream*. For tasks involving repetition, speech information is mapped along a *dorsal stream* from the posterior temporal lobe to the inferior frontal lobe.
Source: Adapted from Hickok and Poeppel (2007).

This second pathway travels from the temporal lobe through superior regions of the frontal lobe associated with the motor system in order to more directly map acoustic input to representations of articulation. This dorsal stream is shown in light gray in Fig. 14.

We'll mostly stick to the ventral path as we move forward to get a handle on how people make meaning out of speech. But as we do this, it will be important to keep in mind two things. The first is that the brain areas we discuss are all parts of a larger multi-faceted network that spans on a great many areas of the cortex. The second point is that the nodes in this network, and hence the observations we make about any specific pattern of brain activity, really depend on what task we've asked the brain to do in a particular experiment.

A neural code for phonemes

The sketch shown at the end of the previous chapter (Fig. 13 on page 60) includes a box labeled *phonological sketch*. What do we know about the neural code for phonological information? The current state of knowledge doesn't have all the answers, and in fact some intriguing clues are not all consistent with each other. I'm going to talk about some of these debates. I also want to show that, despite our somewhat limited knowledge, these neural models are starting to have an impact on our linguistic understanding of phonemes themselves.

A natural starting point is to consider whether the neural code for phonemes, like the neurogram itself, is built on a spatial code. That is, are different phonemes and phonemic features represented by activation of different populations of neurons? Matthias Scharinger and Bill Idsardi (2011) test this proposition by combining the fine spatial and temporal resolution of MEG with the somewhat unique phonological properties of Turkish. The Turkish vowel system is balanced in that it shows a full set of distinctive features spanning front/back articulation, high/low, and lip-roundedness.[1] This linguistic property of Turkish allows the reserchers to test if there is a "neural vowel space" that approximates the acoustic/articulatory vowel space, which is shown in Fig. 15A. They presented Turkish-speaking participants with hundreds of isolated vowels; the participants were asked to press a button whenever they hear a certain consonant among these sounds. These stimuli evoked a brain response in the auditory cortex that peaks about 100 ms after the vowel begins. This is called the "M100" brain response.[2] Because they used MEG, the researchers could estimate the location, or source, of this M100 response fairly accurately. They reasoned that if there were a spatial code for phonemes – just as with neurograms – the location of the M100 peak would differ systematically for different Turkish vowels.

Indeed, Scharinger and colleagues find different source locations for different vowels all along the left superior temporal gyrus, in the vicinity of the auditory cortex. Remarkably, these source locations appear to fall along systematic axes: Front vowels all fall along an inferior-to-superior plane, while back vowels fall along an anterior-to-posterior plane. These axes are shown in Fig. 15B.

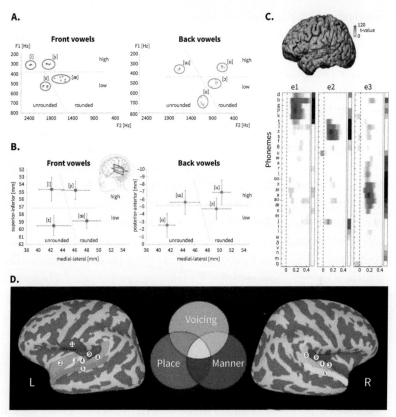

Figure 15. The neural representation of phonemes. (A) Vowels span an acoustic space defined by the first and second formants. (B) Vowels span a neural space in two orthogonal planes around the superior temporal gyrus. (C) Cortical electrodes implanted along the temporal lobe show sensitivity to specific phonological features. (D) The spatial code for phonemic features spans the temporal lobe but not in a clearly defined gradient; non-adjacent regions appear to encode similar information.
Sources: A and B: Scharinger et al. (2011); C: Mesgarani et al. (2014); D: Arsenault and Buchsbaum (2015).

Of course, the vowels do have different acoustics. To test whether these spatial maps follow from the already-familiar spatial code for a neurogram, the researchers considered two statistical models of the spatial location of the vowel. On one model, the spatial distances between vowels should depend on their acoustic differences. On a second model, vowel location was treated as a simple binary difference in vowel features (e.g. back vowel or not? rounded lips or not?). To get the intuition here, consider two specific vowel pairs: the front vowels [i]–[ɛ] and the back vowels [ɯ]–[ɑ]. The acoustic vowel spaces in Fig. 15A show that, within each pair, the vowels differ in height (vowel height information is carried on the first formant). When the acoustics of their stimuli are measured, the back pair differs quite a lot in terms of the first formant (over 300 Hz), but the front pair doesn't differ nearly as much (about 200 Hz). So, if the MEG source data reflects acoustic information, than distance between the pair of back vowels should be greater than the distance between the pair of front vowels. But, if the MEG source data reflect phonemic feature information, not acoustics, then the pairs of vowels should be about equidistant because each pair only differs by one phonemic feature. The statistical models show evidence for features, not acoustics: The spatial locations are about equidistant when vowels differ in just one feature; their distances do not appear to be directly proportional to the acoustic differences. Together, these data appear to indicate that there is a vowel map in the superior temporal gyrus, where distinctive features map to separate axes in space.

Data from other methods support the conclusion that there is a spatial code for phonemic features in the superior temporal gyrus. Should we call this *phonotopy*? In tonotopy, an acoustic dimension (frequency) is mapped to a spatial gradient in the brain; sounds with adjacent frequencies activate neurons that are adjacent to each other. By analogy, the data above seems to suggest that phonemes with similar features activate nearby neuronal populations.

When we expand our view to other kinds of data, however, we do not see strong support for this kind of phonotopy. One such kind consists of ECoG recordings taken directly from the superior temporal cortex while patients listen to naturally spoken sentences (Mesgarani et al., 2014).[3] In fact, we discussed these data on page 53. We observed that parts of the superior temporal gyrus respond selectively to different phonemes, and also that some of those response patterns are categorical, not continuous (we called these *phonemic receptive fields*). But, in contrast to the idea of phonotopy, the researchers did not observe a clear pattern in terms of which recording sites – which populations of neurons – responded to different phonemes. There was no simple "map" on the superior temporal gyrus such that phonemes with similar features showed activation in adjacent recording sites.[4]

A similar, and similarly complex, result has also been observed, using fMRI by Jessica Arsenault and Bradley Buchsbaum (2015). In this study, adults listened to a series of syllables that begin with different consonants. The consonants were chosen specifically to be more or less similar to each other. They measured similarity in terms of phoneme *confusability*: which phonemes are more likely to be mis-heard as each other (so, [t] might be mis-heard as another stop, like [d] or [k], more often than as a fricative like [s]). To identify which brain regions are sensitive to different phonemic features, the researchers used a method for analyzing fMRI data called *multi-voxel pattern analysis*, or MVPA. We haven't seen this before, so I'll take a moment to describe it. MVPA is a useful tool to understand the neural representations that "matter" to a particular set of voxels in the brain. This is a way to ask, now with fMRI, about what the receptive field of a population of neurons might be (but remember that fMRI voxels include tens of thousands of neurons). With MVPA, researchers ask how well voxels in a specific area, called the "searchlight," can distinguish, or classify, different types of stimuli.[5] The basic logic of MVPA is illustrated in the diagram below:

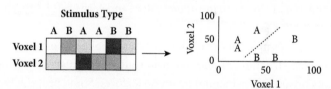

In this diagram the searchlight includes just two voxels. The response pattern for each voxel to two types of stimuli is shown on the left. The plot on the right shows these same activations, now plotted against each other. You can see plainly that the stimuli lead to quite distinct patterns of activation; in other words, these voxels *discriminate* between these kind of stimuli. Equivalently, one can say that the neuronal population in these voxels have receptive fields for whatever feature(s) separate the two stimulus types.

When confronted with different phonemes, regions in the left and right temporal lobe can discriminate them based on phonemic features in a similar way. But, just as with the receptive fields discussed above, the fMRI results did not show clean "phonotopic" maps: Parts of the temporal lobes that are sensitive to the same features are not necessarily next to each other. This is illustrated in Fig. 15D, which shows, for example, that both anterior and posterior parts of the superior temporal gyrus discriminate the voicing feature (in green), but these are separated by regions that also distinguish other features, like the place or manner of articulation.

So, the superior temporal cortex represents phonological detail in a spatial code (*phonemic receptive fields*), but the code does not appear to be organized into a "phonotopic" gradient with similar phonemes activating adjacent neuronal populations.

Neural insight into phonological representations?

What we've learned about the neural code for phonemes may also offer insight into mental representations of phonemes themselves. I'm going to talk about two examples; one is based on data we've seen above and the other gives us a chance to introduce a new, and very useful, experimental tool.

The first example concerns a key question in phonological theory: the nature, and structure, of phonological features. Linguists have proposed various organizing principles for these features. Some propose that features are primarily organized in terms of acoustic properties of speech (e.g. how sonorous different speech sounds might be). Others propose that features are primarily organized in terms of articulation (such as the manner or place of articulation).[6] The ECoG data showing phonemic receptive fields that we discussed on page 53 may shed some light on this debate.

Recall that different electrodes, implanted directly into the cortex, showed responses to specific phonemes. In fact, a close look at the data, illustrated in Fig. 15C, shows that the responses pattern with phonemic *features*, not phonemes themselves; electrodes responded to clusters of phonemes that shared specific features. For example, electrode "e1" in Fig. 15C responds to stops, and electrode "e3" to non-high vowels. The researchers grouped together all the electrodes that showed similar response patterns. This clustering revealed a hierarchical organization of phonemic receptive fields such that one electrode might respond to a subset of the phonemes found for another one. Here's the bit that gets back to mental representations: The highest levels of the hierarchy were split along acoustic features, while articulatory features were only apparent at lower levels of this organization. That is, the hierarchy in the neural code for phonemes seems to be organized primarily by acoustic properties, and secondarily by articulatory properties.

This is an exciting way to use neural data to address questions about mental representations. But using brain data in this way is still a very new idea; the results I just discussed, for example, have only been observed in a just a few datasets from patients undergoing the invasive ECoG procedure. The second example of neural insight into phonological representations is based on an experimental method for studying neural representations that has been used extensively for over thirty years.

People – and their brains – respond quite strongly when they encounter something they see or hear that doesn't match what they expect. But, what counts as a match? For example, if two people with different-sounding voices make the speech sound [ba], does this activate "the same" or "different" mental representations in a listener? Researchers have used the brain's *mismatch response*, or MMR, to probe whether different stimuli match or not to test questions like this. The MMR is observed with electrophysiological techniques like EEG and MEG using an experimental protocol called the *oddball design*. In an oddball experiment, certain stimuli are presented quite frequently – say 80% of the time. These are *standards*. Some stimuli are presented rarely – the remaining 20%. These are *deviants*. The MMR is a signal that emerges around 200 milliseconds after hearing a deviant item, relative to hearing the same stimulus when it is presented as a standard.[7] The diagram below illustrates the oddball experiment:

A as standard: A A A B A B A A A
A as deviant: B A B B B B A B B

You'd expect to see an MMR if you subtract the brain response to the "B" deviant items on the top row from the "B" standard items on the bottom row. By switching out different kind of speech stimuli, we can use this protocol to test questions about phonemic features.

The MMR is sensitive to phonemic differences, not just acoustic differences.[8] That's expected given the evidence we've already seen for the brain's rapid sensitivity (< 200 milliseconds) to phonemic features. Phonologists disagree about the proper structuring of these features, and also about the format used to specify them. For example, is there such a thing as a "default" feature? The intuition behind this relates to efficiency; if phonemes include defaults, then you don't need to store every featural detail separately. Your mental representation only needs to track when a phoneme differs from the default. (The technical term used here is that features may be *underspecified*; Lahiri and Marslen-Wilson, 1991.) One feature that has been proposed to be underspecified is the CORONAL feature that captures the articulation of consonants wth the tongue tip, as with [t], [d], and [s]. This contrasts with sounds like [k] or [g] that are articulated with the base of the tongue, captured with the DORSAL feature. Under this approach, while a [k] is specified with a DORSAL feature, the mental representation of [t] simply lacks any place feature; the system "fills in" CORONAL as the default.

This theory makes a specific prediction about the MMR. Imagine [ka] is used as a standard; this sets up an expectation for more sounds like it. That expectation is mentally represented in terms of features (not just acoustics), so the brain is expecting the next sound to be DORSAL. If the next sound is the CORONAL stop [ta], though, there is a strong mismatch and, consequently, a strong MMR:

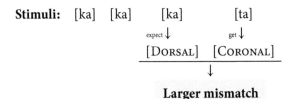

Now consider if [ta] is the standard. If you cash out the expectations this generates in terms of underspecified features, well, there is no explicit expectation for CORONAL (it's the default!). If this is right, then a [ka] deviant will not show as strong a mismatch (remember that the acoustics are different, so there will be at least some mismatch):

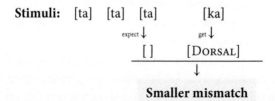

The important point is that this idea of feature underspecification predicts that the MMR is asymmetric: [ta] will be a large mismatch against a standard [ka], but [ka] will be a smaller mismatch against a standard [ta]. Alternatively, if all phonological features are represented equally – no underspecification – then the MMR will be the same regardless of which stimulus is the standard and which is the deviant.

An asymmetry in the MMR is exactly what was observed by Matthias Scharinger, Jonas Obleser, and their colleagues in a study published in 2012.[9] A large MMR was observed in their ERP study for a deviant [ta], while a smaller MMR was observed for deviant [ka]. This result points towards a theory of the mental representation of features that allows some values to be underspecified; there are "default" features.

To summarize this brief section, we've seen two examples of how a better understanding of the brain bases of speech and phonological processing may productively feed back into linguistic efforts to study the mental representations that make up your knowledge of language.

Is speech special? The motor system and perception

We've seen at many levels so far that the brain shows a highly efficient mapping from acoustics to speech. This efficiency is seen already even with classical behavioral findings, such as categorical perception and the McGurk effect. These findings have led to a prominent school of thought which holds that speech perception is "special" in very important ways. This special-ness arises, perhaps, because speech holds a unique role in the acoustic environment of spoken language users.

I want to consider two variants of the *"Speech is special"* hypothesis. One strong version of this view is that speech perception involves perceptual mechanisms that are *different* than the perceptual mechanisms used elsewhere in auditory perception, such as those used to identify a train in the distance, a tea-kettle boiling, or a phone ringing. Such mechanisms could include the unique neural oscillations that serve to sample speech input, or the feedback loops based on higher-level knowledge.

A second, perhaps weaker variant of this hypothesis is that the brain's neural code *is* tuned for speech, but this tuning is built on the same primitive perceptual machinery that is used for other kinds of auditory perception.

I believe that the current evidence best supports the second view, but this debate is far from over, so consider the evidence carefully.

One long-standing piece of evidence pointing towards the "speech is special" idea comes from a peculiar kind of language deficit: *pure word deafness*. In pure word deafness, a patient cannot recognize words, even though they otherwise appear to have normal hearing. Consider the case of patient FO, published in 2017 by Chiara Maffei and colleagues. FO had no difficulty speaking, and her hearing appeared to be intact. For example, she had no difficulty discriminating different vowel sounds based on their acoustics. But she was largely unable to discriminate consonants that were presented as minimal pairs. Remarkably, her performance improved if the consonants were given in writing. Moreover, FO had difficulty discriminating words with similar phonemes in a *picture-naming task*. In this task, the patient might be asked to "point to the *bat*" while viewing a set of pictures that include a bat and also pictures of phonologically similar words, such as a mat. In contrast, FO had no difficulty when distinguishing words that had similar meanings ("Point to the *table*" alongside pictures of a table and a chair), nor did she have such difficulty when working with written words.

The pure word deafness apparent in patients like FO contrasts with another deficit, *auditory agnosia*. *Agnosia* is a general term for a deficit associated with recognizing objects. So, auditory agnosia is a more specific deficit associated with recognizing objects based on their sound. A patient with auditory agnosia will, for example, have difficulty recognizing the roar of a lion or the notes of a piano. Such a pattern was observed in a case study of patient LD published in 1989 by Jany Lambert and colleagues. In addition to being unable to name familiar sounds, LD was unable to recognize well-known folk songs, and mistakenly identified music or environmental sounds as speech. LD was somewhat impaired in recognizing phonemes and words, but this impairment was notably milder than the profound deficit they had in recognizing sounds, easily recognizing vowels and scoring near perfectly when confronted with phonological neighboring words.

Pure word deafness and auditory agnosia, together, present a remarkable double dissociation:

	FO	LD
Recognizing speech (pure word deafness)	*Impaired*	*Not impaired*
Recognizing non-speech sounds (auditory agnosia)	*Not impaired*	*Impaired*

This double dissociation indicates that, at some important level, the cortical processing of speech is separate from the cortical processing for other sounds.[10]

What sets speech apart from other kinds of auditory input? One notable fact is that we are generally expert *producers* of speech in addition to being expert *comprehenders*. If our speech production expertise aids in speech perception, this could explain what sets speech apart from other kinds of audition. The idea here is that brain systems involved in articulation, including our motor system, may be central to the specialness of speech.

In fact, we've already seen a clue that points towards a role for the speech production. Look back at Fig. 13 on page 60. This figure sketched the mapping of acoustic input, via a neurogram, to phonemes. It includes a box called "analysis-by-synthesis." This box aims to capture how we form predictions about speech input that help to guide our perception. Page 59 introduced the idea that predictions come from the listener's ability to internally produce speech sounds, to *synthesize* speech—that internal synthesis can be compared from the "top-down" against auditory input from the "bottom-up".

There is some intriguing neural evidence for such a motor pathway in speech perception. In one study, Friedemann Pulvermüller and colleagues (2006) engaged participants in several speech tasks during fMRI scanning. In the first task, participants produced [pa] and [ta] syllables. Recall from Chapter 2 (page 21) that the primary motor cortex is spatially organized so that the region that controls the tongue is separated from regions that control other parts of the face, such as the lips. Accordingly, Pulvermüller et al. observe separate patterns of activation along the motor cortex when participants produce a bilabial [pa] syllable (closure of the lips) compared to an alveolar [ta] syllable where closure is made by the tongue tip (go ahead: make these sounds yourself!) Next, participants simply listened to those same two types of sounds. In addition to activation in the auditory cortex, researchers also saw activation in the motor cortex when participants simply listened. And, this activation appeared to be spatially separated according to articulators: The tongue region activated positively for [ta] but not [pa], while the lip region activated for [pa] but not [ta].

Moreover, there is evidence that this motor cortex activation is causally connected to successful speech perception. Riikka Möttönen and Kate Watkins tested this in a 2009 study by temporarilly inhibiting motor cortex activation using the rTMS technique (see page 39 in Chapter 2). Möttönen and Watkins reasoned that if the motor cortex is actively involved in speech perception, than discriminating between phonemes that differ in one feature (say, place of articulation) will be more difficult if activation in the motor cortex region associated with articulating those phonemes is inhibited.

Indeed, when the rTMS inhibition targeted the lip region, participants showed a reliable decline in their ability to distinguish bilabial [pa] or [ba] segments from alveolar [ta] or [da] segments. Careful control conditions demonstrated that this decline in perceptual acuity was specific to inhibition of the lip motor cortex itself: rTMS did not impact their ability to discriminate phonemic pairs that didn't involve the lip articulators (e.g. [ka] vs [ga]). Further, when rTMS was applied to other parts of the motor cortex, like the hand region, there was no measurable effect on speech perception.

It appears, then, that brain regions involved in controlling articulation may also play an active role when simply listening to speech.

How crucial is this motor system connection for successful speech perception? Some researchers have proposed that the motor system is absolutely central for comprehension and, perhaps, for other aspects of language processing as well.[11] However, there is quite a bit of evidence that speech perception – though it may *make use* of the motor system – does not *depend* on that system for success. One such piece of evidence comes from aphasia patients with preserved speech perception despite severe deficits in speech production. This is, in fact, the classical characterization of "Broca's" aphasia; see Table 2 on page 13. Corianne Rogalsky and Gregory Hickok (2011) described five compelling case studies showing that successful speech perception does not depend on intact speech production abilities. All of the patients suffered from left hemisphere lesions that severely impacted speech fluency. Two of the patients had lesions that extended across the motor cortex and to more anterior regions of the frontal cortex also associated with controlling actions like articulation. Despite severe disfluency, all participants performed well on a series of speech comprehension tasks. All of the patients showed high accuracy when discriminating words or syllables that formed minimal pairs. For example, in a picture-naming task (see page 72), the participants with frontal damage only – including the motor cortex – performed perfectly in identifying the correct word when there was a phonemically related distractor (e.g. "point to the coat" alongside a picture of a goat).

One possible critique against lesion-based evidence like this, though, is that the brain may reorganize to compensate for neural trauma. Neural plasticity remains poorly understood, and this uncertainty encourages us to look for possible corroborating evidence. A striking source of further support comes from an invasive "brain anesthesia" procedure used in some neurosurgeries. In the *Wada procedure*, an anesthetic agent (usually sodium amobarbital) is injected into a major artery serving one hemisphere of the brain. This, essentially, puts that side of the brain to sleep. The patient is then engaged in a series of language tasks. If the patient is unable to produce speech during the procedure, the affected hemisphere is said to be "dominant" for language.[12] What happens to speech comprehension in this procedure when production is completely impaired? The patients made very few phonological comprehension errors, assessed using a picture-naming task similar to the one just described; their average accuracy was greater than 90% (Hickok et al., 2008). These data indicate that motor systems involved in speech production aren't necessary for successful comprehension.

The picture emerging from the data so far is that the motor system is engaged during speech perception but is not necessary for success. One way to look at it is that the expert knowledge we possess as producers of speech offers a helpful boost when faced with particularly difficult input. Such a supporting role is consistent with the idea from Fig. 13 that top-down knowledge refines our analysis of the speech signal.

Let's take a very brief look at a second possible specialization for speech: temporal windows of integration. As discussed around page 55, it seems as if auditory regions of the superior temporal gyrus sample speech in a precise way that seems tuned to the specific properties of speech. For example, the more rapid 25–40 ms sampling rate is suitable for analyzing the fine structure of speech, while the longer 200–300 ms rate is suitable for syllabic information. Is this tuning special to speech? Well, first, it's not actually quite clear whether these neural oscillations are tuned for speech, or whether speech itself is tuned to fit in with intrinsic properties of our auditory system. Setting aside this "chicken-and-egg" challenge, there is good evidence that other sensory inputs are sampled in the way we've suggested for speech.

In a groundbreaking 1993 study, Stanislas Dehaene observed that behavioral reaction times, the speed at which a participant is able to press a button in response to a stimulus, follow a periodic pattern; participants responded at certain regular intervals. This observation came from over 1,500 reaction times collected from participants who performed a set of both auditory and visual detection tasks (for example, "Did the letter T or L appear on the screen?" Or, "Did you hear a high note or low note?".) Across tasks and participants, the reaction times were periodic. The exact timing of these periodicities differed across tasks. For more difficult tasks this periodicity was at a rate of about 33 Hz or, equivalently, reaction times were at intervals of about 30 milliseconds. (Shorter intervals, as fast as 10 milliseconds, were observed for easier tasks.) On the basis of these data, Dehaene suggested that perceptual systems in the brain pass on information to higher-level processing at fixed intervals, or windows. There is evidence, then, from non-linguistic tasks using both visual and auditory stimuli, that perception more generally might operate according to fixed temporal windows of integration.

In this section we've touched on the fundamental question of whether "speech is special". We've seen evidence that, indeed, speech comprehension relies on neural systems that are distinct from those used for non-speech perception (e.g. pure word deafness), and such systems may include the motor/articulatory system used for speech production (e.g. inhibiting motor cortex with rTMS). But those special speech production systems do not seem to be necessary for speech comprehension (shown, for example, by patients undergoing the Wada procedure). And the apparent tuning of auditory processing to speech sounds seems to be an instantiation of more general sensory processing systems that operate over fixed temporal windows. Speech perception is special in how finely tuned it is, but this tuning is built on top of general purpose neural mechanisms for perception. Summarizing their own research on this topic, Han Gyol Yi, Matthew Leonard, and Edward Chang put the point this way (2019, p. 1099):

> the cortical infrastructure for auditory processing [is not] entirely specific nor selective to speech [...] but is nevertheless heavily specialized and causal for speech perception.

Insight into speech perception disorders?

What we've learned so far about how the brain processes speech can provide insight into speech disorders. I won't review speech comprehension disorders thoroughly here, though recall we've been introduced to some already in this chapter, like pure word deafness. What I want to do instead is show one example of how our growing understanding of the brain bases of speech may translate into more concrete clinical applications.

Autism Spectrum Disorders (ASD) are a complex multi-faceted developmental syndrome affecting about about 1 in 54 children in the United States (Maenner, 2020). Impaired communication, including language abilities, is one of several core symptomologies in ASD. Outcomes span a very broad range from children who are non-verbal to high-functioning individuals with few if any language-related difficulties.[13] Alongside language-related deficits are differences in sensory perceptions.[14] A major open question in ASD research is whether, and in what way, lower-level sensory differences connect with higher-level issues affecting communication and other social behaviors. I'm going to discuss a few intriguing lines of research probing the neural link between auditory processing and speech perception in ASD.

One piece of neural evidence for sensory differences in ASD comes from the auditory M100 brain response that is measured with MEG. Recall that the M100 is a MEG signal from the auditory cortex that peaks about 100 milliseconds after the presentation of a sound; it reflects a relatively early stage of auditory integration (see page 64). This response appears to be delayed in children with ASD. For example, a 2010 study by Timothy Roberts and colleagues at the Children's Hospital of Philadelphia studied a sample of children around 10 years old with and without an ASD diagnosis. They observed that the M100 MEG response was 10 or so milliseconds later, on average, in children with an ASD diagnosis compared to their peers. In follow-up work, this delayed brain response to auditory stimuli was linked to differences in neural oscillations, specifically oscillations between about 20–50 Hz. Differences in the strength of these oscillations *prior* to a simple acoustic stimulus correlate with delayed M100 latencies in a sample of 105 school-aged children with an ASD diagnosis (Edgar et al., 2015). This is the same neural oscillation pattern that has been linked to the mapping from continuous acoustic input to categorical phonological representations (see page 55).

Could altered neural oscillations associated with mapping from acoustic to phonological representations contribute to language-related deficits in ASD? The jury is definitely still out on this question, but there is a growing body of evidence suggesting that such a link is plausible.[15] In the study of 105 children mentioned above, for example, the strength of neural oscillations correlated with a behavioral assessment of language abilities. A 2015 study by Delphine Jochaut, Anne-Lise Giraud, and colleagues makes the case for an even tighter link between language deficits and atypical auditory-to-phonology mapping. They combined spatially precise fMRI with EEG data to assess how neural oscillations from the auditory cortex support speech comprehension in ASD. They make three key observations. First, auditory cortex activity in children with ASD shows reduced sensitivity to the speech envelope, compared to activity from age-matched peers (recall Fig. 11). Second, this reduced envelope-tracking is linked with neural oscillations, as children with an ASD diagnosis show atypical low-frequency oscillations in a band of 4–7 Hz – this is the same band discussed on page 55 that is associated with tracking temporal properties of speech, like the envelope. Altered 4–7 Hz oscillations are further linked with altered higher-frequency oscillations, from 30–40 Hz (associated with fine structure and phonemic analysis). These researchers suggest that atypical speech envelope-tracking may affect the brain's ability to "lock on" to the speech signal to efficiently extract phonemic details. Indeed, these atypical oscillations correlate with, for example, verbal communication scores in that sample of children.

There are a number of important questions to be asked of this line of research. Some intriguing findings, like the link between low- and high-frequency oscillations in ASD, are based on a relatively small number of individuals. Even for results from larger samples, the degree of heterogeneity in language and other communication abilities in ASD can make it difficult to draw broad generalizations. One specific challenge comes from observations that there may be multiple different kinds of language impairment sub-types that co-occur with ASD (Rapin et al., 2009). Keeping these questions in mind, the key take-away here is the hypothesis that temporal sampling of speech, and therefore phonological processing, may be impaired in ASD. Importantly, this research has been guided by our growing understanding, from linguistics and neuroscience, of the neural mechanisms that map from continuous acoustic input to categorical linguistic mental representations.

Chapter summary

Let's summarize what we've seen both in this chapter and in the previous one by revisiting the three transformations that the brain carries out to convert acoustic input into linguistic representations:

acoustic signal →₁ neuro-auditory representation (*neurogram*) →₂ neuro-phonologcal representation (*phonological sketch*) →₃ lexical item

We've learned some key facts about the brain bases of each of these transformations:

→ 1 Acoustic information is converted to a neural representation of sound, a *neurogram* in the primary auditory cortex within about 100 milliseconds.

These representations encode both the fine structure spectral information that distinguishes different phonemes, and also the temporal envelope information that encodes syllabic information. These details are represented with a *spatial code*: Different neurons respond to distinct spectral and temporal information.

→ 2 *Neurograms* are used to activate neural representations of phonological features. Populations of neurons adjacent to the auditory cortex show *categorical* responses to features within 100–200 milliseconds after speech is heard.

This mapping may be carried out in at least two separate *temporal windows* that serve to take "snap shots" of the speech input: a shorter window, around 25 milliseconds, tuned to phonemic features, and a longer window, around 200 milliseconds, tuned to syllabic features.

This mapping follows a feedback loop: Linguistic knowledge helps to *synthesize*, or predict, upcoming speech input which serves to guide and refine the *analysis* of that input. While this process is helped by specialized linguistic knowledge, it seems to rely on the same basic neural architecture used for other kinds of auditory input.

→ 3 The neural representation of speech provides clues to the mental representation of phonological information in the mental lexicon. We saw how data like the spatial organization of *phonemic receptive fields*, the M100 evoked response, and the *mismatch response* can be used to test specific theories about phonemic features. While preliminary, neural evidence points towards a priority for acoustic features, and these features may be *underspecified*.

These neural signals indicate that speech perception is built on general auditory processing systems that are highly tuned to the specific properties of speech. Ongoing work is probing how aspects of this tuning play a role in developmental disorders such as *Autism Spectrum Disorder*.

How the brain uses this information to access and understand words is the main topic of the next chapter.

5
Activating words

The next two chapters address the neural representation of words and concepts. What is a word? Perhaps surprisingly, it has proven difficult to find a satisfying scientific definition for this term. The working definition I'll use for this chapter is that a word is a pairing of linguistic *form* with *meaning* and *structure*. When we talk about this pairing as a mental representation, something you hold in your mental dictionary, I'll call it a *lexical item*. The form is made up of phonological sequences. Those phonological representations may come from speech, as we studied in the previous chapter, or sign languages. The form may also include written (*orthographic*) components. The meaning, or semantics, includes the concept that the word refers to as well as other semantic features. The structure indicates how the word fits together with others into a grammatical sentence. The pieces of this definition fit together like so:

$$\text{word} := \text{lexical item} := \langle \text{FORM}, \text{MEANING}, \text{STRUCTURE} \rangle$$

where FORM includes phonology, orthography... and MEANING includes concept, semantic features...

So, we can write out a lexical item for, say, the word "cat" roughly like this: </kæt/, 🐈, NOUN>.

This definition is a bit of a departure from what you might call a "word" in day-to-day usage. For example, a *morpheme*, like the plural suffix "–s" at the end of "cats," expresses phonological form, meaning, and structure, and so it meets this definition. Morphemes are, indeed, lexical items that cannot be divided into smaller meaningful units. But, our definition of lexical items allows for them be made up of multiple morphemes; "classroom" could be a single lexical item, for example. Indeed, one of our tasks in this chapter will be to examine the "grain size" of lexical items: are they stored in the brain as single morphemes, or can they be chunked together into complex combinatoric units? Alongside this question, we'll examine the brain systems involved in accessing these units. The next chapter then dives into a series of debates involving the mental representation of concepts; we'll see here how brain data offer new insights into some old philosophical questions about meaning.

Words and Wernicke

The evidence suggests that linguistic form and meaning are combined together – made into lexical items – in the *posterior middle temporal gyrus*, or pMTG. This node forms a kind of interface between linguistic features, like phonology, and more general aspects of conceptual semantics. I like to think of the lexical items in this node as a sort of contact card, a mental entry that links the non-linguistic concept, such as my memory of the soft, purring, attention-demanding pet I have, with a linguistic name: /kæt/.

Posterior areas of the left temporal lobe, especially the middle temporal gyrus, have long been linked to lexical processing and semantics. This connection goes all the way back to the 1870s and Carl Wernicke who, as we saw in Chapter 1, documented how damage to posterior parts of the temporal lobe led to patients showing a distorted kind of semantics; they produce semantically inappropriate words and also show great difficulty with understanding.

Modern deficit/lesion studies support the link first documented by Wernicke. A particular striking example comes from a large-scale study of 101 aphasia patients. To take advantage of this unusually large dataset, Elizabeth Bates, Nina Dronkers, and colleagues (2003) used a variant of the *lesion overlap* method discussed in Chapter 2 on page 30. With all the datasets in their sample, they formed two groups based on whether or not each patient had a lesion in a particular brain voxel (Fig. 16, top). Then, they tested whether or not the two groups differed on a clincial measure, such as semantic comprehension or speech fluency. They repeated this procedure for many different voxels across the brain. This procedure is called *voxel-based lesion symptom mapping*, or VLSM.

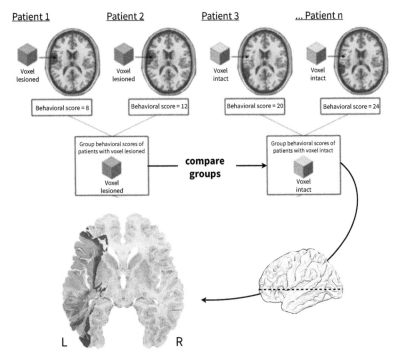

Figure 16. Voxel-based lesion symptom mapping (VLSM). Patients are grouped by whether or not they have a lesion at a particular voxel, and their clinical performance is compared (top). This yields a map that is shaded to indicate which locations are most associated with a particular performance deficit. The bottom shows such a map for semantic comprehension, as assessed via a picture-naming task. Shades of red indicate voxels in the left posterior temporal lobe that are strongly associated with deficits in this task, while shades of green and blue indicate left hemisphere voxels where damage is not strongly associated with task performance.
Sources: Top: Adapted from Baldo et al. (2012); Bottom: Adapted from Bates et al. (2003).

The results of the VLSM procedure is a map showing which voxels are systematically linked with a particular behavioral deficit. One such map is shown on the bottom of Fig. 16.

They observed, like Wernicke almost 130 years prior, that patients with damage in the posterior area of the left temporal lobe, especially the pMTG, showed the most reliable difficulties with language understanding, such as picture-naming tasks where the patient follows an instruction like "point to the table." These patients had relatively fluent speech production, indicating that any difficulties with comprehension were not based on lower-level phonological deficits. In other words, the *form* part of their mental representations for words was intact. The difficulty with word-understanding documented here could be due to two different kinds of semantic deficits. They could have a deficit in relation to lexical items themselves, or, alternatively, there could be deeper deficit in relation to conceptual processing.

Evidence points to the first hypothesis: Patients with pMTG damage have relatively intact conceptual networks, but the link between those concepts and lingusitic form – the lexical item – has been impaired. One piece of supporting evidence is that patients with comprehension deficits like these demonstrate implicit conceptual knowledge (so-called "Wernicke's aphasia"; Table 2 on page 13). *Semantic priming* is one tool to measure implicit conceptual relationships. The basic idea is that people are quicker and more accurate in accessing a conceptual representation if they've just processed something that is semantically similar. A common way to test this is to present people with word pairs and ask them to judge if all the items are real words ("doctor", "glove"...) or not ("soam", "flipo"...). People are faster at this *lexical decision task* if the word pairs are related ("doctor", "nurse") than if they are not ("bread", "nurse").[1] This speed-up reflects the fact that the concepts which the words refer to are related; when you mentally access one concept, it's easier to activate other concepts that are "in the same neighborhood."

Patients with fluent aphasia seem to show this same semantic priming speed-up (Milberg and Blumstein, 1981). These patients are generally much slower than non-aphasic individuals in performing the lexical decision task, but they nevertheless show a speed-up, measuring about 1/5th of a second, when words are conceptually related to each other, compared to when they are not. This result provides evidence that these patients still have intact conceptual representations, representations that capture the relationships between similar concepts like "tree" and "leaf." Accordingly, the semantic deficit in these patients appears to be in linking up phonological form with the correct conceptual meaning – a deficit with the lexical item itself.

Measurements of brain activity in response to real words and nonwords also highlight the role of the pMTG. Matthew Davis and Ingrid Johnsrude (2003) offer one example of this in an fMRI study where participants listened to sentences that were distorted in different ways so as to make them more difficult to understand. These distortions included adding noise to the stimuli, replacing small segments of stimuli with bursts of noise, or using vocoding to remove the speech fine structure (see page 47 in Chapter 3). They reasoned that the intelligibility of the stimuli would affect brain areas associated with lower-level auditory and phonological processing as well as higher-level lexical processing. Among these brain areas, lower-level regions will be particularly sensitive to the specific form of distortion used, while areas associated with higher-level comprehension processes will not be sensitive to the specific kind of distortion used. They observe "form-dependent" brain activation around the auditory cortex and adjacent regions on the superior temporal gyrus. In contrast, they see "form-independent" activations across the pMTG and also more anterior areas of the temporal lobe.

We'll come back shortly to what those anterior regions might be involved in. But here, we see further evidence linking the posterior middle temporal gyrus with a stage connecting speech form with more abstract meaning.

Studies like this one point towards a hierarchy of processing in the brain, as information flows from core areas that capture sensory details onwards to higher-level, more abstract representations, first phonemes and then lexical items and concepts. We saw in the last chapter how the early stages of this flow feed into different pathways, processing streams, for high-level processing; check out, for example, Fig. 14 on page 63. In the next section, we'll see how tools like MEG, with their high temporal resolution, let us follow this cascade of information in real time.

The time-course of word recognition

In 2003, Ksenija Marinkovic and her colleagues at Massachussetts General Hospital offered a simple, yet stunning, view of the cascade of brain activity that goes into recognizing words. Using MEG, they recorded brain responses from participants who either listened to, or read, single words. To ensure participants considered the meaning of the words, they had to judge whether they referred to things that were large or small. Because the magnetic fields measured by MEG are relatively undistorted as they pass through the skull, the source of brain activity can be reconstructed with reasonable accuracy. When these reconstructions apply to the entire cortex, they form a kind of "brain movie" that shows the dynamic ebb and flow of brain activity over space and time. Still images from the brain movies created by Marinkovic and colleagues are shown in Fig. 17A.

The top row of Fig. 17A shows the response for spoken words. Just 50 milliseconds after a spoken word is presented, activity begins to emerge in the primary auditory cortex. Just 120 milliseconds later, less than 1/5th of a second, that activity has spread to adjacent areas of the superior temporal gyrus. This is the time-window in which acoustic information is being mapping to phonological representations. 250–300 milliseconds after the word began, activation has spread outwards along the temporal lobe, including posterior areas of the middle temporal gyrus associated with lexical processing, and also anterior areas that we have yet to discuss. Simultaneously, activity emerges in the inferior frontal gyrus. This is another key node in the network of regions processing higher-level aspects of language processing; we saw it already in our discussion of Paul Broca on page 10 in Chapter 1 and also in Fig. 14 on page 63. The functions carried out in these frontal regions is the subject of a major debate, but we're going to hold off on engaging with it now, and keep our focus on the cascade of processing going on in the temporal lobe.

The bottom row of Fig. 17A shows the brain activity that is observed for written words. This cascade begins not in primary auditory regions, but rather in primary visual cortex located in the occipital lobe. As with

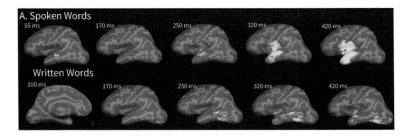

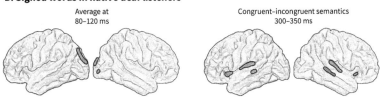

Figure 17. The time-course and localization of lexical activation.
(A) Activation of left-hemisphere brain activity in response to spoken (top) and written (bottom) words. Activity starts in sensory cortices (auditory or visual) and then moves to shared higher-level regions by about 250 milliseconds after word onset. (B) The time-course of brain activity for signed word recognition begins in the primary visual cortex (left) but rapidly spreads to the same higher-level temporal and frontal brain regions within about 300 milliseconds.
Sources: A: Adapted from Marinkovic et al. (2003, fig. 1); B: Adapted from Leonard et al. (2012, fig. 2).

spoken words, activity spreads from core sensory areas to higher-level areas; for visual information, this spreads along the so-called "ventral pathway" that is involved in recognizing objects.[2] 170 milliseconds after appearing on the screen, the word is activating a ventral region at the border of the temporal and occipital lobes; in the left hemisphere this brain region seems to be involved specifically in recognizing letters, earning the nickname of the *visual word-form area*. Up to this point, the pathway for spoken words and written words has been very distinct. Keep in mind also that written words are presented to the visual system all at once, in contrast to spoken (and signed) words, which unfold over time. Despite these striking differences, their paths begin to merge within just 200 milliseconds in the posterior temporal lobe, especially the pMTG. After about 300 milliseconds, activation continues to spread to anterior temporal and inferior frontal regions; these temporal and frontal regions are involved in high-level aspects of word processing regardless of whether the input was written or spoken.

A very similar picture of the time-course, and relevant brain regions, emerges from studies of word recognition in signed languages. An example is shown in Fig. 17B. As with written-word recognition, signs first evoke visual activity in the occipital lobe, which moves to the same higher-level frontal and temporal lobe sites seen for spoken language within about 300 milliseconds.[3]

The cognitive processes that contribute to this complex pattern of brain activity are just beginning to be understood. One open question concerns the degree of back-and-forth interactive processing between higher-level frontal and temporal regions and lower-level regions associated with sensory processing like the auditory cortex. While the picture presented in Fig. 17 appears to show mostly "feed-forward" activations, this may reflect a limitation of studies that use isolated words, as opposed to linguistic input that comes from a richer and more natural context. We saw this issue already come up when examining how predictions help in speech perception (see around page 59), but the core idea is really quite general; indeed, one leading idea is that the brain ought to be fundamentally understood, at a much more general scale, as a "prediction engine."[4]

A great deal of research aims to unpack the different stages of processing implicated in this cascade, and how these processing stages might interact. But there are some fundamental properties of language that have made these efforts challenging. Chief among them is the fact that linguistic computations and representations are, themselves, interrelated in complex ways. To give one example of these challenges, I'd like to talk about a study that seems to contradict some of the conclusions about the neural time-course of word recognition that we've reached so far. In this study, led by Lucy MacGregor of Cambridge University (2012), participants listened to real English words and fake words while MEG was being recorded. The fake words were very carefully constructed so that they sounded as similar to real words as possible (pronounceable fake words are called *pseudowords*). For example, real words in their study included "beak" ([bik]), "moat" ([mot]), and "ripe" ([raɪp]), while pseudowords included "bik" ([bɪk]), "mort" ([mɔrt]), and "roop" ([rup]). Not only were the pseudowords carefully matched in pronounceability, but they were constructed so that the *uniqueness point* – the moment that the stimulus becomes recognizable as not real – was timed precisely. In analyzing the MEG data, the researchers tested when differences in the neural response emerged after this uniqueness point. Remarkably, they found differences in the brain response as early as 50 milliseconds after this uniqueness point. As we discussed in Chapter 3 (see page 45), this is about the amount of time it takes for information to pass from the cochlea to the cortex. It seems from these data that lexical access – distinguishing real words from pseudowords – happens as soon as auditory information arrives in the cortex.

Should we reconsider our models of speech perception to allow for lexical access in 50 milliseconds? There may be another explanation for those results, and this explanation speaks to how inter-related linguistic representations make it difficult to tease apart different stages of processing in the brain. One such difficulty is that phonological sequences which belong to lexical items are more frequent, and therefore more predictable, than phonological sequences which don't belong to a lexical item. Indeed, there *must* be some correlation between phonological sequence frequency and lexical frequency, because we (typically) only come across phonological sequences when they make up words. With this in mind, we can see at least one other interpretation of the extremely fast brain response to pseudowords discussed above: The response at 50 milliseconds could reflect a neural "error signal" when encountering unexpected sensory input. For example, given a word beginning [bɪ], the brain might then expect [t], [n], or [b], which come from words like "bit", "bin", or "bib".[5] Indeed, we saw evidence that predictions like this play an important role in the brain's capacity to comprehend variable and noisy input back in Chapter 3 (see page 58 and also Fig. 13). If the next sound is instead [k], then there is a sensory mismatch between your phonological predictions and the actual input:

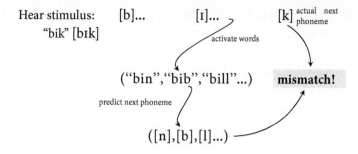

A mismatch like this, between your expectations and the form-properties of the input, could lead to a relatively rapid brain response that correlates with whether the input is or is not a real word.

Is there evidence that early brain responses associated with sensory input can be shaped by phonological predictions? Some interesting data come from experiments using visually presented words, rather than auditory speech. Written words are useful in this case because early sensory responses to visual input in the occipital lobe can be easily distinguished from higher-level phonological and lexical regions in the temporal lobe. In 2009 and 2010, Suzanne Dikker, Hugh Rabagliati, and colleagues at New York University designed a series of studies in which participants read words that did or did not fit with the sentence they appeared in while recording with MEG.[6] Words that did not fit had the wrong *syntactic category* (e.g. noun, verb, adjective...):

> *match:* The tasteless soda...
> *mismatch:* The tastelessly soda...

In these examples, the noun "soda" is expected after an adjective like "tasteless", but a noun is not grammatical if it appears after the adverb "tastelessly". And indeed, the researchers see an increase in early activity in the occipital visual cortex for the unexpected noun. This visual response indicates that expectations for the *form* of a particular word may influence the earliest stages of sensory processing in the cortex. Form-based expectations like this may also explain the apparent extremely rapid response for pseudowords, mentioned above.

The broader take-home message is that linguistic form properties, like whether phonological sequences are more or less predictable, can be very difficult to fully separate from lexical properties.

Although different levels of linguistic representation are inter-related, they can be teased apart under certain carefully constructed circumstances. While frequent words will generally have more predictable phonological sequences, the exact values of these two measures aren't identical; common sequences may also appear in rare words. The quantitative differences between form properties and lexical properties are lost when words are grouped into categories – experimental conditions – but they can be separated when the data are analyzed at the level of the individual experimental stimulus. This is called a *single-trial analysis*. Gwyneth Lewis and David Poeppel use this approach in a 2014 study designed to tease apart early form-processing stages from later lexical processing stages.[7]

In this study, participants listened to hundreds of monosyllabic words during MEG scanning. The words were chosen to span a range of form-based and lexical variables. The researchers then analyzed the data by looking for neural activity that correlated uniquely with each of the variables. Phoneme predictability, measured by the frequency of phoneme pairs (called *bigrams*), correlated with activity 100–200 milliseconds after word onset in the left superior temporal gyrus right around the auditory cortex. Following this form-based response, brain activity correlated with lexical measures. One such lexical measure is the size of the *lexical cohort*; this is the number of words in your mental dictionary that begin with the same set of phonemes (so, the cohort of /pi/ includes "pea", "peach", "peanut", "peak"...). Notice that the brain needs to start accessing the mental lexicon in order to take stock of a word's cohort size; activity consistent with the initial activation of a lexical item emerged around 250 millseconds in the superior temporal sulcus, posterior to the auditory cortex. The frequency of a word itself, independent of these other factors, correlated with brain activity in a later window, 300–400 milliseconds. By this stage, representing access of the target lexical item, activity had moved to the posterior middle temporal gyrus. Lexical activity in the middle temporal gyrus in this study nicely parallels the evidence from deficit/lesion studies, discussed around page 84 above, that damage to this same region impairs access to lexical items.

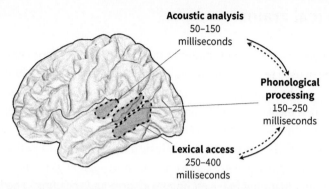

Figure 18. Stages of spoken-word recognition. Spoken-word recognition unfolds in at least three stages, all within less than half a second. Acoustic analysis starts in the primary auditory cortex, followed by phonological processing in adjacent parts of the superior temporal gyrus, and finally lexical access in the pMTG. Dashed lines indicate that higher-level information feeds back to facilitate lower-level processing in a dynamic process that is not yet well understood.
Source: Adapted from Lewis and Poeppel (2014).

The illustration in Fig. 18 summarizes the sequence of neural processing stages that is beginning to emerge from this research on the time-course of spoken word recognition. Acoustic and phonological analysis takes place in and adjacent to the auditory cortex within the first 150 to 200 milliseconds after a word is presented. This information is then passed to posterior middle temporal regions where to activate lexical items, completing the mapping from linguistic form to meaning.

There is a whole host of open questions we can ask at this point. We've seen that lexical information can influence earlier sensory processes – how are these feed back loops neurally implemented? What about feedback from higher levels of linguistic processing (sentences, discourse, etc.)? We'll come back to some of the complex interactions between feedback and feed-forward processing in Chapter 7. For now, let's turn to another central question for the neurolinguistics of words: What is the unit of lexical representation?

Lexical grains

What we've learned thus far about the time-course of word recognition has been used to shed light on the nature of lexical representations themselves. At the beginning of this chapter we defined the term *lexical item* to mean a mental representation that maps between phonological form and meaning. Let's use *mental lexicon* to refer to the part of our memory that stores these representations. A lexical item that cannot be broken up into smaller parts that are also form–meaning mappings is a *morpheme*: a minimal mapping between linguistic form and meaning. The brain's lexical items could correspond to a morpheme in a one-to-one way, or maybe the brain has lexical items that correspond to words made up of multiple morphemes. The word "cat" has one morpheme (it is *monomorphemic*). It seems intuitive to say that "cat" is also a lexical item. The word "cats," with the plural "-s" suffix, has two morphemes. Is "cats" a single lexical item in the lexicon, or do we recognize this word by accessing two different mental representations: "cat" and "-s"?

The question becomes perhaps more interesting when we consider a broader range of words (and a broader range of languages, of course, although I'm going to stick with English for my examples here). What about an irregular plural word, like "geese"? Like "cats" it has two pieces of meaning (an animal piece and a plural piece); but do we say that the change in vowel ("oo" → "ee") is, itself, a lexical item? Similar complications arise from other types of words; take the word "natural," which has two morphemes, the noun "nature" and the suffix "-al," which makes the word an adjective.[8] Do we have one word "natural" in our mental lexicon, or do we store "nature" and then, separately, store a lexical item for "-al"? Does "naturalize" activate three separate lexical items?

All of these questions come down to the unit, or "grain size", of the lexical items that are stored in the brain. Do we represent lexical items as minimal units – as morphemes? The theory that we do so is called the *full decomposition theory* of lexical access. As such, this theory requires the brain to decompose an input like "cats" or "natural" into constituent parts, access each lexical item separately, and then recombine the pieces into larger units according to the rules of grammar.

Table 4. Two theories for how lexical items are represented

Word and morphemes	Lexical items according to... the *full decomposition* or...	the *partial decomposition* theory
"cats" cat+s	</kæt/, 🐱 > </s/, PLURAL >	</kæt/, 🐱> </s/, PLURAL >
"geese" goose+[plural]	</gus/, 🦢> <ø, PLURAL>	</gis/, 🦢🦢>
"natural" nature+al	</nætʃɹ/, 🐦> </ʌl/, ADJECTIVE>	</nætʃɹʌl/, ADJECTIVE-OF- 🐦>

An alternative theory holds that we sometimes store lexical items built from multiple morphemes: irregular words like "geese" or words like "natural" correspond to a single whole lexical item. This second view is called *partial decomposition*. Table 4 summarizes these two theories.

Neurolinguists have approached these questions using a variety of methods. Interestingly, and puzzlingly, different tools seem to yield results that support different conclusions. On the one hand are deficit/lesion studies showing a double dissociation between regular and irregular morphology. Stephen Pinker and Michael Ullman (2002) point to series of case studies defending the partial decomposition view.[9] One patient, JLU, had *anomia*. This is a kind of aphasia that leads to difficulty finding the right words – think of words being forever "on the tip of your tongue." JLU had special difficulty with irregular words, like using "held" as the past tense of the verb "hold." In contrast to JLU, patient FCL had a kind of aphasia called *agrammaticism*. This is a variant of non-fluent aphasia (see page 13) where patients have special difficulty with function words, words that carry syntactic information, as opposed to content words. Patient FCL performed poorly when generating regular past-tense words like "walked" or "jumped," and in fact they performed relatively better in producing irregular verbs like "held" or "ate." Here is the familiar pattern of a double dissociation.

	FCL	JLU
Regular verbs ("walked," "jumped")	*Impaired*	*Less impaired*
Irregular verbs ("held," "ate")	*Less impaired*	*Impaired*

The researchers suggested that double dissociations like this one support the partial decomposition viewpoint. Regular words like "walked" (and "cats" etc.) are composed following grammatical rules; when these rules are impared in patients with agrammaticism, so the theory goes, patients will have specific trouble with the regular words that follow these rules. On the other hand are irregular words like "held" (also "geese" etc.). On the partial decomposition view, these words are stored as whole units, not made up of pieces that combine together. These whole units are not impaired in agrammaticism, but they may be impaired in patients with anomia, who have difficulty specifically with accessing words.

A different picture has begun to emerge, however, from experiments that look at the time-course of word recognition for complex words. In one such study in 2015, Laura Gwilliams and Alec Marantz of New York University probe Arabic word recognition using MEG.[10] This is a good reason they chose to study Arabic in this study. Whereas complex words in English are built up by adding suffixes or prefixes, Arabic, like other Semitic languages, forms certain complex words by inserting different vowel patterns between consonants that form the word root. Here is a simplified example:

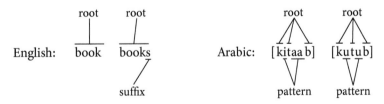

So, morphemes in Arabic can be interwoven with each other; they are "non-linear." This property of Arabic allowed the researchers to tease apart two aspects of accessing morphemes that are otherwise highly correlated: the probability of the morpheme you might next encounter, and the probability of the phoneme you might next encounter. In a language with linear morphology, like English, these two probabilities tend to go hand in hand. If you're thinking that this reminds you of something, it should. The correlation between morphology and phonology is cousin to the correlation between lexical properties and form properties that we discussed earlier in the chapter (see page 91). Just as we saw above, morphological factors can be teased apart from phonological factors by examining how they vary across individual words – individual trials – in the experiment. Specifically, the researchers quantify the probability of encountering the last consonant in each Arabic word in two different ways. A *phonological* probability is conditioned on just the previous phonemes that have been encountered. A *morphological* probability is conditioned on the separation of the word into distinct morphemes.

How does this relate to lexical grain size and full or partial decomposition? Well, the full-decomposition theory asserts that the input word must be decomposed into minimal morphemes before it can be accessed. So, morphological structure should affect relatively early stages of processing, prior to 250 milliseconds when lexical access begins (see Fig. 18). In contrast, the partial-decomposition theory asserts that words can be accessed as whole units, so there is no expectation that internal morphological structure should guide early stages of word recognition. Indeed, morphological structure appears to guide processing of spoken Arabic words even in early stages, between 100 and 200 milliseconds. In fact, the researchers did not see in their data any reliable effects of the linear phonological factor at early or later processing stages. The results suggest that spoken-word recognition, which involves using predictions as a word unfolds, relies on decomposing the input into component morphemes prior to accessing the meaning of the word.

What are we to make of the apparently conflicting findings about the grain size of lexical items? On the one hand are double dissociations in aphasia which seem to indicate that complex words can be stored and accessed as whole units, but on the other hand, evidence from MEG suggests that words are decomposed into their constituent morphemes at early stages of processing, even when the morphemes are non-linear.

When you confront apparent contradictions like this, there are two general strategies for moving forward. You can think carefully about the *evidence*: Does it mean what you really think it means? You can also think carefully about the *theories* at issue: Are there aspects of the theories that, when they are drawn out and made more clear, can start to help sort out different kinds of data?

I think it's illustrative to briefly consider both strategies when we think about lexical grain sizes. (This is far from the last unsettled debate we'll see in the book. We'll certainly revisit this exercise again.) From the perspective of the evidence, we can see that the data that I've presented here actually look like a comparison of apples to oranges. I mean something deeper than observing just that deficit/lesion correlations and MEG data are different. Rather, the way they differ, and the kind of brain function they can reveal, can give us important clues as to how to reconcile these findings. For example, the deficit/lesion data provide no window into the time-course of word recognition. So, if irregular words like "geese" or "held" involve different kinds of processing at any stage – early *or* late – then this could lead to the double dissociation that we saw above on page 95. This leads to a question we might ask of the theories we have been considering: If lexical items are fully decomposed in the way illustrated in Table 4, then the brain still must have some process to handle irregularities in order to, for example, change the vowel from "goose" to "geese." If we consider those processes carefully, could they perhaps help explain how an account of full decomposition can also capture the double dissociation observed in aphasia?

There are a number of other threads we might pull on to help reconcile these two competing accounts (Data: Have double dissociations like those found with English words also been found with Arabic words? Theory: For partial decomposition, what determines when words are stored as wholes or in parts; could Arabic non-linear morphology actually be the latter? And so on...). What I want to leave you with here is that our growing understanding of how the brain recognizes words, the brain areas involved, and their time-course, is opening up new windows into fundamental questions like what is the mental representation of a "word"? Progress on answering these questions comes both from looking at a broader range of evidence and from sharpening our theories of what an adequate answer might be.

Chapter summary

This chapter has tackled how the brain recognizes words.

- Words, or more accurately *lexical items*, are made up of phonological, semantic, and structural features. Word recognition requires mapping a phonological representation to a mental representation of meaning.
- This mapping happens rapidly in the left temporal lobe of the brain, as phonological representations in the *superior temporal gyrus* are used to activate lexical items in the *posterior middle temporal gyrus*.
- Evidence from MEG outlines an extremely rapid sequence of processing stages, from acoustic analysis at 50–100 milliseconds to phonological processing at around 100–150 milliseconds, and onwards to lexical access in the posterior middle temporal gyrus after just 250 milliseconds, or 1/4th of a second.
- Ongoing work is trying to uncover the lexical units that form the bases for recognition; at least some data support the *full-decomposition theory* that lexical items are recognized after first breaking them into their component morphemes. Of particular interest here are apparent conflicts between data from different methodologies; several possible strategies are available for reconciling these findings.

If lexical items map from form to meaning, then the natural next question is this: How does the brain represent meaning?

6
Representing meaning

The striking thing about words – or, more precisely, lexical items – is that they mean something. Indeed, the nature of this meaning has been an enduring puzzle facing generations of philosophers, linguistics, psychologists, and now neuroscientists. This chapter introduces some of the insights into this challenging question that have emerged from studies of the human brain.

Distributed and non-distributed conceptual representations

From the earliest accounts to the current state of the art, neuroscientists have agreed that "meaning" only emerges when vast parts of the brain work together. Here is Carl Wernicke, writing on the topic in 1874:[1]

> the memory images of a bell [...] are deposited in the cortex and located according to the sensory organs. These would then include the acoustic imagery aroused by the sound of the bell, visual imagery established by means of form and color, tactile imagery acquired by cutaneous sensation, and finally, motor imagery gained by exploratory movements of the fingers and eyes[.]

The leading idea here is that the concept "bell" is made of many different kinds of knowledge, including memories of how bells sound, look, feel, and how you use them. The different kinds of knowledge that go into a concept become even more complex for feelings like "happy" or abstract ideas, like "fair" or "enough." The discipline that focuses on word meanings is *lexical semantics*; this is not a book about that very rich topic. I'm going to focus on just two aspects of conceptual meaning that connect with our understanding of the brain. Both of these are already hinted at in the quotation from Wernicke. The first is the idea that conceptual meaning is *distributed* throughout the brain. In particular, we'll examine whether such distributed representations are sufficent for capturing how the brain represents concepts. The second idea, which we take up in the next section, is the degree to which conceptual meaning is *embodied* in our perceptual and action systems (are concepts "deposited [...] according to the sensory organs"?), or whether meaning is more *abstract*.

There is a rich body of evidence supporting the idea that conceptual meaning is, indeed, distributed across different cortical systems. One kind of evidence comes from a remarkable pattern of deficits called *category-specific agnosia*. I introduced agnosia first in Chapter 2 (page 29; also page 72) as a deficit in recognizing objects or concepts. Patients with category-specific agnosia have this difficulty with certain specific classes of objects. Many such cases have been carefully documented by Elizabeth Warrington and her colleagues at University College London.[2] A picture-naming task is one useful tool to test for category-specific deficits (see page 72). One patient, named SBY, performed pretty well identifying pictures of household objects and other inanimate items, both when the item names were spoken and when they were written down. But, when presented with pictures of living things, like animals, SBY could not correctly identify a single one. Strikingly, SBY's difficulty didn't seem to relate to the complexity or abstractness of the concepts being tested. When asked to give definitions for words, SBY did quite well with a variety of objects or even quite abstract ideas, as shown in the examples on the left below, in contrast to the living things shown on the right:

submarine "ship that goes underneath sea"
wheelbarrow "object used [...] to take material about"
malice "to show bad will between people"
caution "to be careful how you do something"

wasp "bird that flies"
duck "an animal"
frog "an animal, not trained"
tobacco "one of the foods you can eat"

Deficits like that of SBY have been documented in many patients. Some show an even more remarkable degree of specificity. For example, one case study identified a person with difficulty recognizing fruits and vegetables, but not animals, tools, or vehicles (Samson and Pillon, 2003)!

Which brain regions are involved in representing these different kinds of concepts? In 2009, Jeffrey Binder of the Medical College of Wisconson and colleagues conducted a *meta-analysis* of 120 fMRI studies addressing where conceptual meaning is stored in the brain. A meta-analysis pools together the quantitative results from multiple studies to ask which patterns generalize across different experiments. A challenge for any meta-analysis is determining when different studies are addressing "the same thing" in the brain. In this analysis, three different kinds of experiments were pooled together: Experiments comparing real words with pseudowords, experiments where participants focused on meaning or focused on phonological form, and experiments comparing semantic meaningfulness (e.g. whether words are semantically related to each other or not). Their results are summarized in Fig. 19A. This summary makes clear that, as Wernicke originally surmised, brain areas across the temporal, frontal, and parietal lobes are involved in processing conceptual meaning. Binder and colleagues further examine studies that probed specific aspects of semantics, such as those that probe for "action" meanings, or for the meanings of man-made objects (10 studies each). These more specific comparisons reveal subparts of this broader network, including posterior parts of the temporal lobe that border on the parietal lobe (specifically, the *angular gyrus*). The activation patterns for different types of meaning are distinct, consistent with the occurrence of category-specific agnosia, discussed above.

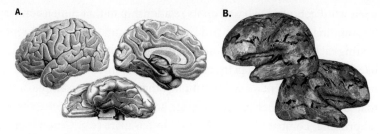

Figure 19. The semantic system. (A) Regions associated with the brain's distributed "semantic system" are found in the temporal, frontal, and parietal lobes in a meta-analysis of 120 studies. (B) A semantic atlas for a single-subject implicates a wide range of regions; even words with similar meanings, like the "social" words in red, activate a broadly distributed network.
Sources: A: Binder et al. (2009); B: Adapted from Huth et al. (2016).

A stunning example of how rich semantic representations are distributed across the cortex comes from a study by Alexander Huth, Jack Gallant, and colleagues at the University of California, Berkeley (2016). What sets this study apart from prior research, including the meta-analysis just mentioned, is that they were able to map semantic networks for many different kinds of meanings simultaneously. To do this, they had participants listen to spoken stories during fMRI scanning. These naturalistic stimuli had richer semantic content than typically found in highly controlled experiments. To determine how the brain responds to this rich semantic content, the researchers created a mathematical representation of the meaning of each word. Such a representation, called a *word embedding*, places each word in a "semantic space" whose dimensions reflect different kinds of meanings. The following diagram, for example, includes three dimensions: "food," "movement," and "space":

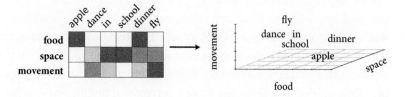

Each word gets a value based on its relationship to that dimension (say, by counting how often the word "apple" co-occurs with a word representing each of the dimensions in a large corpus of text). Semantic spaces like the one schematized in the above diagram seem to capture aspects of how people represent meaning. Among other things, for example, words that are close together in such a space are more likely to cause semantic priming, compared to words that are far apart (e.g. Günther et al., 2016). Having creating such a semantic space, the researchers then use statistical models to estimate which voxels in the fMRI images respond selectively to certain dimensions of meaning in this space.[3] This procedure reveals a quite stunning "semantic atlas" showing how different kinds of meanings activate different areas of the brain. An example of this atlas for a single participant is shown in Fig. 19B. The colors in this figure represent different dimensions in the semantic space. For example, voxels shown in red are more sensitive to words that have a social meaning ("family," "parents," "met"), while areas colored in green relate to words with more visual meaning ("yellow," "stripes," "shaped").

There is a lot of detail in this analysis and in the results that we don't have space for here. I want to highlight just two takeaways. First, we see again that different brain regions respond more to certain kinds of meanings than others. This matches what we already saw in the studies discussed above. Second, and adding new complexity, we see that similar types of meaning are themselves distributed across brain regions; for example, you can see brain areas responding to social aspects of meaning in the temporal lobe and frontal lobe in both hemispheres (red areas on Fig. 19B).

Mapping out where different sorts of meanings are represented in the brain can offer a window into questions about the mental representation of meaning itself.

One approach builds on the computational models of lexical semantics used by Alexander Huth and colleagues in the study just described. Rather than build just one model, or word embedding, of semantics, they actually created models using several different techniques. To understand this, imagine taking the diagram on page 103 and changing the labels on the plot. Different models correspond to different choices about what the dimensions – the axes on the plot – ought to be. They find that a model whose dimensions are based on a vocabulary of basic words provides a better match of brain activity than a model based on more abstract word-to-word co-occurrences. This line of research is just beginning, and there remain many unanswered questions about what sorts of semantic spaces best represent human conceptual knowledge, and how those semantic spaces are physically instantiated by distributed neural networks.

Another question about semantic representations concerns whether the distributed networks we've discussed so far are *sufficient* for capturing how the brain represents conceptual knowledge. The *distributed-only theory* holds that concepts emerge solely from this distributed network. This theory is schematized at the top of Fig. 20A. An alternative theory is that the distributed aspects of conceptual meaning must be bound together in some way to form a coherent concept that is abstracted away from particular experiences. This alternative *distributed-plus-hub theory* is illustrated at the bottom of Fig 20A.

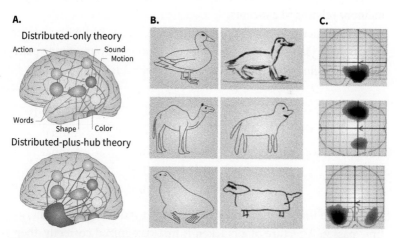

Figure 20. Semantic dementia. (A) Illustration of the distributed-only and distributed-plus-hub hypotheses for the neural representation of meaning. (B) Examples from a *delayed-copy task* indicate that the loss of conceptual specificity in semantic dementia is not unique to language. (C) The anterior temporal lobes are the main site of neural degeneration in semantic dementia.
Source: Patterson et al. (2007).

The hub in Fig 20A covers the anterior temporal lobe. You may recall the anterior temporal lobe from the previous chapter, around page 88, where we saw that this region shows a systematic response to a semantically meaningful stimulus around the same time as the posterior middle temporal gyrus that houses lexical items. Researchers have proposed that the anterior temporal lobe may be a crucial "semantic hub" that binds together the distributed aspects of conceptual meaning.

One central piece of data pointing towards an anterior temporal hub for semantics is a remarkable language deficit called *semantic dementia*.[4] Semantic dementia is a kind of *primary progressive aphasia*, or PPA. That means it is a language deficit that follows from a progressive neurodegenerative disorder. This contrasts with the aphasia that we have seen previously, where neural damage is caused by external factors like a stroke or traumatic brain injury. Like the more familiar neurodegenerative disorder Alzheimer's Disease, semantic dementia affects memory with progressively severe consequences. Whereas Alzheimer's affects *episodic memory* – the memory of specific events, people, and places – semantic dementia affects semantic knowledge, or memory relating to concepts.

The disorder in semantic dementia can be illustrated with a picture-naming task (see page 72). Examples of responses from a single patient are shown below ('+' indicates a correct response):

	Responses from			
	Sept. 1991	Mar. 1992	Sept. 1992	Mar. 1993
Bird	+	+	+	Animal
Chicken	+	+	Bird	Animal
Duck	+	Bird	Bird	Dog
Swan	+	Bird	Bird	Animal

Notice here that the responses appear to get more vague or generic as time passes. A picture of a chicken is first recognized correctly, then is only recognized as a "bird", and finally only as an "animal."

Other tests show that semantic dementia really affects the concepts that are connected to words, not to the lexical item itself. One task that shows this non-linguistic aspect of the deficit is called the *delayed-copy task*. This task is illustrated in Fig. 20B. Patients are shown a picture (left-hand column of the figure). The picture is then taken away and the patient is asked to draw the picture from memory. Examples of drawings from a semantic dementia patient are shown on the right-hand side of Fig. 20B. You can see that the drawings are not perfect reproductions of the original picture. Interestingly, the drawings seem to be lacking the unique and specific features that set each animal apart: The camel lacks a hump, the seal has gained legs and a tail, and the duck has four legs instead of two. The drawings seem to resemble what you might get if you asked someone to draw "an animal"; just like their language, the drawings are more generic. Evidence from tasks like these indicate that semantic dementia impacts conceptual knowledge in a general way as patients lose the ability to recognize or reason with the features that distinguish different concepts from each other.

While the illustrations I've given here relate to animal concepts, the effects of semantic dementia are seen across many different conceptual categories. But, despite the category-general nature of this deficit, semantic dementia does *not* affect a broad, distributed, neural network. Rather, semantic dementia is associated with the progressive degeneration of neurons in a rather focused location: the anterior temporal lobes in both hemispheres. Fig. 20C illustrates the anterior temporal brain areas that show reduced functioning in semantic dementia.

The fact that damage to a relatively focal region of the cortex leads to a general semantic deficit is not predicted under the *distributed-only theory* that was introduced above. Under that view, damage to a focal part of the semantic system should only affect certain kinds of knowledge. Rather, the deficit observed in semantic dementia supports the *distributed-plus-hub theory*; the anterior temporal lobes are, under this view, the hub in which different kinds of conceptual knowledge are bound together.[5]

Evidence from other methods also supports the distributed-plus-hub view. Repetitive TMS, or rTMS, can be used to selectively disrupt different parts of the semantic network (see page 39 in Chapter 2 for a refresher). When rTMS is applied to the anterior temporal lobe, it seems to induce a brief category-general deficit, but when it is applied to another node in the semantic network, a category-specific deficit is found. To show this, Gorana Pobric, Elizabeth Jeffries, and Matthew Lambon-Ralph (2007) used rTMS to disrupt three specific nodes: the left anterior temporal lobe, the inferior parietal lobe adjacent to the motor cortex, and a control region in the occipital lobe that is not expected to disrupt any semantic processing. Participants were then asked to make semantic judgments about several kinds of stimuli: high- and low- action pictures that were expected to be associated with the motor cortex, and pictures of living and non-living objects. Number words were included as a control condition, as they are not expected to engage the conceptual system. Indeed, when rTMS was applied to the inferior parietal lobe, participants were slower to make decisions about which words were related to actions, but were not affected in making decisions about whether a word named a living or non-living thing. This is a kind of category-specific effect. But when rTMS was applied to the anterior temporal lobe, participants were slower making both kinds of semantic judgments in a more category-general way. The overall pattern of results can be summarized:

	rTMS applied to		
	Anterior temporal hub region	Inferior parietal motor region	Occipital control region
High/low action	*slower*	*slower*	–
Living/non-living	*slower*	–	–
Number words	–	–	–
	↓	↓	
	category-general	category-specific	

These data from healthy participants appear to line up with the category-general pattern of deficits in semantic dementia patients, consistent with the theory that the anterior temporal lobe is a kind of semantic hub.

One last bit of data supporting the *distributed-plus-hub* view comes from fMRI. Jenny Crinion, Cathy Price, and their colleagues at University College London (2006), studied the brain bases of semantic processing in bilingual participants. They were especially interested in what sorts of brain areas support semantic processing that is *common* across languages, versus what sorts of brain regions might support *different* languages. They use a variant of semantic priming (see page 85) called *fMRI adaptation*. The idea here is that stimuli that engage the same neuronal populations – say, two words that activate the same semantic representations – will lead to reduced activation after repeated exposure. Accordingly, stimulus items could be semantically related or not, and they could be drawn from the same language or from different languages. Here is a schematic of the experimental setup for participants who spoke English and German ("forelle" is the German word for "trout"):

		Prime word	Target word
related	*same language*	trout	SALMON
unrelated	*same language*	trout	HORSE
related	*different languages*	forelle	SALMON
unrelated	*different languages*	forelle	HORSE

Words that were semantically related led to faster responses – semantic priming – and this was true even if the words came from different languages. Moreover, just one brain region showed this priming pattern for semantic relatedness across languages: the anterior temporal lobe in the left hemisphere. Other experiments probing bilingual semantic representations also point to the anterior temporal lobe. For example, the pattern of activation in the left anterior temporal lobe observed for words in one language can be used with the MVPA technique (page 67) to predict when participants were presented with a word with the same meaning, but in a different language (Correia et al., 2013). It appears that lexical items from different languages point to a shared neural representations of meaning in the anterior temporal lobe.

These bilingual data, like the rTMS and semantic dementia data discussed above, are consistent with the theory that the anterior temporal lobe is a semantic hub. This area serves to represent conceptual information that is abstracted away from specific sensory or motoric modalities. This hub is shared across different languages, and is used for semantic processing in linguistic and in non-linguistic tasks.

Are conceptual representations embodied?

By this point, we've seen quite a bit of evidence that the brain represents concepts through a complex network of brain regions distributed around the cortex, and it seems that these representations are bound together in a semantic hub housed in the anterior temporal lobe. Given the evidence for a semantic hub, let's revisit the role that distributed activations play in instatiating conceptual knowledge. We'll consider two possible hypotheses:

> **Grounded symbolic concepts** are built from sensory and action experiences and are bound into abstract units in the anterior temporal lobe. Distributed activation in sensory or motor systems is not necessary for the representation of the concept once it is formed.
>
> **Embodied concepts** are built from sensory and action experiences, and those same systems are necessary for the representation and processing of those concepts. For example, understanding an action-word like *walk* requires activating the same motor representations used when performing that action.

The debate between these two hypotheses is not yet settled, and we will see data supporting both sides in this section. The balance of the evidence at this point seems to me to favor the *grounded symbolic concepts* perspective, but this again is another opportunity to consider how to reconcile apparently conflicting findings in neurolinguistics.

Carl Wernicke, if you remember from the evocative quotation on page 100, suggested that conceptual knowledge relies on brain regions involved in sensory perception and even "motor imagery gained by exploratory movements of the fingers and eyes." The *embodied concepts* theory predicts that the motor system is necessary to understand words related to performing actions. Indeed, the involvement of motor systems in representing concepts seems to receive striking support from a number of studies using fMRI and TMS.

In one such study, Friedemann Pulvermüller and colleagues had participants perform a series of simple actions while undergoing fMRI scanning, like wiggling their fingers, moving their feet, or puckering their lips. You might recall from Chapter 2 that the motor cortex of the frontal lobe is organized such that different cortical areas control different parts of the body (page 21). Indeed, when participants moved an arm, leg, or their face, the researchers saw activation in the corresponding areas of the motor cortex. Then, the researchers simply had the participants read words that were projected onto a screen. Among the items that were presented were action words, like "kick," which involved moving the legs, or "throw," which involved moving the arm, or "kiss," which involves the face. According to the *embodied concepts* theory, a concept like "kick" involves, in part, memories based in the motor/action system itself. And indeed, the researchers in this study saw activation in the motor cortex as participants simply viewed action words on the screen. Moreover, the activation pattern, while not identical to the pattern that was observed when they had performed related actions, showed a similar spatial distribution: Leg-related actions like "kick" activated a region similar to actually moving the foot, arm-related words activated a region similar to actually moving the arm, and so forth. Data like this support the idea that conceptual representations involve brain regions that are engaged when we interact with and perceive things related to that concept.

Under the embodied view, our ability to comprehend action words should be impaired if the relevant parts of the motor system are disrupted. Indeed, we already saw above some evidence suggesting that this might be the case. In the study by Gorana Pobric and colleagues (2007) discussed on page 107, participants showed a category-specific slowdown when making judgments about pictures of action-related items when rTMS was applied to the inferior parietal lobe, adjacent to the somatosensory and motor cortices.

Another study that indicates a possible causal connection between motor cortex activation and action semantics used TMS to stimulate, not disrupt, motor cortex processing (Pulvermüller et al., 2005). Whereas repeated pulses of magnetic stimulation from rTMS briefly inhibit neural activation, a single magnetic pulse can enhance neuronal excitability within a small area of the cortex. In this study, such single pulses were applied selectively to the arm-area or the leg-area of the motor cortex.

Control conditions involved TMS pulses to the right hemisphere. Participants made lexical decision judgments about a series of words with meanings related to actions involving the arms ("fold," "grasp") or legs ("kick," "hike").

The results showed that participants were faster at making lexical decisions when the semantics of a word matched the brain site that was stimulated. So, when stimulation was applied to the arm-region of the motor cortex, participants were about 15 milliseconds faster in making a decision about arm-related words like "grasp." But, when stimulation was applied to the leg-related region, participants were instead faster to make judgments about leg-related words like "hike" by about 30 milliseconds.

These TMS studies indicate that the comprehension of action words can be briefly impaired or enhanced by stimulating specific areas of the motor cortex. It's important to keep in mind what we mean by "impaired" and "enhanced" here – the impact of TMS on participants is being measured in milliseconds, such that a participant might be a few tens of milliseconds slower or faster in making a judgment about a word than they otherwise would be. Evidence like this seems to suggest that a more severe disruption to the motor system ought to lead to a more severe impairment with processing action-related meanings. However, severe disruptions to the motor system of the brain do not, it appears, have equally severe consequences for the comprehension of action-related concepts; this is at odds with the *embodied concepts* view, but it is consistent with the *grounded symbolic concepts* theory.

Parkinson's disease is a neurodegenerative disorder that is specific to neurons in the motor system. As a consequence of this neurodegeneration, patients with Parkinson's disease lose aspects of motor control, leading to tremors and to becoming slower with controlled actions like walking. The *embodied concepts* theory suggests that patients with severely disrupted motor systems will also have difficulty with language comprehension involving action-related words. David Kemmerer of Purdue University and colleagues tested this prediction in a study 2013. They probed semantic comprehension in a sample of patients with Parkinson's disease as well as control participants of the same age.

None of the participants showed evidence of dementia. Semantic comprehension was tested by having participants make a *semantic similarity judgment*:

"Which word is most similar to TRUDGE?"

LIMP   STROLL

You can see that this kind of judgment requires relatively subtle semantic knowledge. Participants made judgments like this about various kinds of action words as well as non-action words that describe mental states ("happy," "afraid," etc.) Although the semantic judgments were subtle, participants with Parkinson's disease performed very well regardless of whether the judgments involved different kinds of action words or non-action words. There were no observable differenes in the semantic performance between participants with and without Parkinson's disease. This result is not consistent with the *embodied concepts* theory.

More evidence that difficulty performing actions doesn't correlate with difficulty comprehending action concepts comes from case studies of *apraxia*. Apraxia is a deficit involving performing actions due to brain damage. Rafaella Rumiati, Tim Shallice, and colleagues (2001) document a series of case studies that show a double dissociation between apraxia affecting certain types of actions and understanding of those actions.[6] Patient DR and patient FG both experienced strokes that damaged significant portions of their left hemisphere, including the motor cortex. Both of them showed evidence of apraxia: They had dificuly imitating actions that were demonstrated to them, and also had difficulty demonstrating how objects might be used ("Show me how you use a phone"). These patients were then given an assessment designed to test whether they understood the actions that they could not perform. They were asked to arrange into the proper order sets of pictures corresponding to the steps of some action. Both patients performed well at this picture-sequencing task, even though neither one could perform even half of the actions that were pictured. Moreover, the same sequencing task was given to another stroke patient, WH2, who showed no difficulty with action performance – no apraxia – but rather showed executive functioning difficulty. While WH2 could perform the depicted actions, they had relatively greater difficulty putting the action pictures in the correct sequence:

	DR & FG	WH2
Performing pictured actions	*Impaired*	*Less impaired*
Sequencing pictures of actions	*Less impaired*	*Impaired*

These case studies demonstrate a double dissociation between being able to perform an action, affected in apraxia, and the semantic comprehension of the action. This double dissociation is not consistent by the *embodied concepts* theory.

Let's take a moment to consider how these different pieces of data fit together to shed light on theories of conceptual representation.[7] On the one hand is evidence showing that distributed parts of the brain, like the motor cortex, become activated even when participants simply comprehend lexical items. Moreover, there is some sort of causal link between motor activation and semantics, as individuals are measurably faster or slower to recognize action words, specifically, depending on whether the motor cortex is stimulated or inhibited. The data appear to support the claim that the motor cortex is *involved* in aspects of semantic processing, but they are not alone sufficient to support the stronger claim of the *embodied concepts* theory that the motor/action system is *necessary* for semantic comprehension. Instead, evidence from damage to the motor system indicates that the ability to perform an action can dissociate from the capacity to understand that action and words related to actions. The balance of the evidence seems to point to the *grounded symbolic concepts* theory of an action system that is connected to conceptual representations, but is not necessary for semantic comprehension of actions themselves.

Chapter summary

This chapter has tackled how the brain represents word meaning itself.

- A broadly distributed network of cortical regions in the temporal, frontal, and parietal lobes are activated when a person performs the incredibly complex task of semantic processing. State-of-the-art

methods are beginning to uncover the atlas of semantic information that is encoded across these *distributed* semantic networks.
- Distributed semantic information appears to be bound together to form coherent concepts in a *semantic hub* brain region: the anterior temporal lobes. Key evidence for this hub comes from *semantic dementia*, which is a variant of *primary progressive aphasia* (*PPA*).
- While semantic representations are distributed in the brain, it is not as clear that those representations are embodied in sensory and action systems, following the *embodied concepts* hypothesis. For example, there appears to be a dissociation between the ability to perform actions and the capacity to understand action-related semantics. This dissociation supports the *grounded symbolic concepts* hypothesis.

Aside from the specific findings and theories discussed in this chapter, we also spent some time thinking carefully about what to do with experimental results that appear to conflict with each other. The discussions in the last few chapters of the time-course of word recognition (page 87), lexical grains (page 94), distributed concepts (page 100), and embodied concepts (page 110) all asked us to confront evidence from studies that didn't seem to line up. I hope you see in each of these sections that such conflicts are scientifically good! They require us to think carefully and skeptically about what the data at hand really mean, and to make our theoretical questions and hypotheses sharper and more effective.

While these last two chapters have covered quite a lot of ground, you won't be surprised to learn that we've just touched the surface of the many rich lines of research that are delving deeper into how the brain represents words and meanings. I encourage you to explore the notes sprinkled through the chapters! Or, if you're ready, you can turn to the next chapter, where we try to make sense of how the brain puts words together to make meaningful phrases and sentences.

7
Structure and prediction

Everyone creates brand new sentences every day. I'm not talking about poetry or word play here, though there is that too, but just your everyday bit of communication, like "Don't step on that bear!" or "The fox that was under the bed is now in the kitchen." Now I suspect that you might not have come across these particular sentences before. My point is that you had no problem, no problem at all, understanding what each of these mean. Having reached that understanding, you may have even done some further reasoning ("Hmmm, it seems pretty implausible to step on a bear. Maybe the word *bear* refers not to a large carnivorous mammal, but rather to a soft plush toy"). To do this, you took the words that you know, and put their meanings together in a rule-governed and systematic way to come to an understanding of the meaning of the whole sentence. The next three chapters are about how the brain carries out this remarkable feat of composition.

Sentence structure

Though it seems effortless to us, there is a lot that goes into making meaningful sentences out of strings of words. Fig. 21 shows a phrase-structure diagram that aims to capture just some of the pieces that we need to keep in mind when we think about what the brain must be doing. There are quite a few things going on this diagram, but I want you to focus on just three key aspects of sentence structure:

Constituency Words combine together in a particular order. This is indicated by the hierarchical groupings, called the *constituency* or *phrase structure* of the sentence. So, the word *red* combines with *apple*, and that pair of words together combine with *the*. Together, these make a phrase: *[the [red apple]]*. Similarly the phrase *[this*

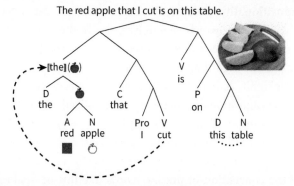

Figure 21. Components of sentence structure. Words are organized into hierarchical phrases, or constituents, governed in part by their syntactic category. Dependencies between words may span long distances, as indicated by the curved lines. This structure, together with the words themselves, compositionally determine the meaning of the whole sentence.
Source: Apple image by Marco Verch, used under the CC BY 2.0 license (https://creativecommons.org/licenses/by/2.0/)

table] combines with the preposition *on* to make the phrase *[on [this table]]*. Constituency plays an important role in indicating, for instance that *I* is the subject, and agent, of the verb *cut*. This information about "who did what to whom" is the *argument structure* of the sentence.

The rules for constituency are based on each word's *syntactic category* (or *part of speech*). These are the "P", "N", or "V" labels in the figure that indicate whether a word is a preposition, noun, verb, etc. There are some labels that may be unfamiliar ("C"? "Pro"?). Don't worry about these. Any abbreviations will be defined as needed through the chapter.

Dependencies Sometimes words combine with elements that aren't right next to them. These are called *long-distance dependencies*. This is indicated by the dashed arrow connecting the phrase *[the red apple]* with the verb *cut*: this line captures the fact that it's the apple that is being cut.

Another kind of dependency is when words must show *agreement* in their form. This is indicated in the dotted line in the diagram, which shows how the demonstrative word *this* agrees with the noun *table*. If the noun was plural (*tables*), the demonstrative would instead be *these*.

Compositionality Finally, these phrases and dependencies, together with the words themselves, derive the meaning of an expression. That meaning is indicated, very roughly, by the picture on the right-hand side of Fig. 21 Importantly, this meaning is built *compositionally* from the pieces of the sentence. This compositional process is schematized by the small pictures corresponding to *red* (■), *apple* (☼), and phrases like *[red apple]* (●). Some words don't have meanings that are easily shown with an icon or picture, such as the word *the*; I'll use the convention of putting words like this in double brackets (⟦•⟧) when referring to that word's meaning.

Where do these structures come from? Well, one thing that any language user knows (though not necessarily consciously) are what sorts of phrasal constituents and dependencies follow the rules of their language. We call this knowledge the *grammar* of a language. Of course, whether we are signing, speaking, or reading, we don't come across a fully-formed sentence all at once. Rather, we perceive a sentence one word at a time. The brain's job is to dynamically make use of this grammatical knowledge to "figure out" the sentence's structure – its constituency, dependencies, and ultimately the meaning – from a sequence of words. (You may recall that we first brushed up against this way back on page 7 in Chapter 1.) So, the task faced by the brain is how to "figure out" which grammar rules match some sequence of words so that it can compose word-meanings together in the correct way to reach some intended meaning. In comprehension, this figuring-out of sentence structure is called *parsing*. When we measure brain signals, then, we're not measuring the brain basis of a grammar directly, but rather we measure the reflexes of this parsing process:

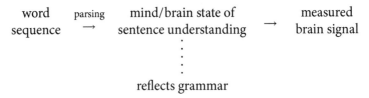

The relationship between the grammar and the brain signals we measure is thus not direct; it depends on several *linking functions*. These are the arrows in the diagram just above. They describe how some linguistic input maps to mental representations, how those mental representations rely on your grammatical knowledge, and how our measuring tools tap into those mental states.

We'll have to do some work in this chapter to understand these linking functions. For a start, it's helpful to think about this parsing process, this "figuring-out" of sentence structure, as a kind of memory game. That grammatical knowledge you need is stowed away in your memory. The game is to dig into your memory to find and apply just the bits and pieces needed to make sense how the words you encounter fit together. You win the game when you understand some linguistic input, and you get points for doing this very quickly (remember, natural speech unfolds at two to six words per second!). To do this effectively, the brain makes predictions about what sorts of grammar rules – what sorts of structures – might best fit a sentence as it unfolds.

The Ns and the Ps of sentence predictions

In previous chapters, we saw how predictions help to guide speech perception of variable input (page 58), and aid in extremely rapid word recognition (page 89). The structural dependencies between words make predictive processing an especially important tool for rapidly and efficiently processing sentence meaning. We'll see, in fact, that predictions play a role in almost every aspect of how the brain makes sense of sentences.

To begin, we'll discuss one of the best-studied brain signals associated with sentences, a signal called the "N400." The N400 is an *event-related potential*, or ERP, component, that is measured with EEG (need a quick ERP refresher? See page 35 in Chapter 2). What this means is that it is a systematic voltage fluctuation that occurs every time the brain encounters a certain type of stimulus. In this case, the N400 is evoked in response to semantically meaningful stimuli. It is measured as a *negative* voltage fluctuation around the center of the scalp, hence the "N" label, that occurs between around 300 and 500 milliseconds after a meaningful stimulus; it is strongest at around 400 milliseconds, leading to the "400" label.

Marta Kutas and Steven Hillyard first measured the N400 in a study published in 1980. In that study, they used EEG to measure how the brain responds to different kinds of unexpected stimuli. In one condition, they varied whether the final word of a visually presented sentence appeared in the same font as the previous words. When visual expectations were violated, this led to a large positive voltage; this is shown in the dotted line in Fig. 22A. (Note that positive voltages are plotted *down* in Fig. 22; this is a common convention in ERP plots but it is not always followed. My advice is to always pay close attention to the axis labels.) However, when the final word of the sentence appeared in an expected font, but had an unexpected meaning, then there was a large negative voltage, which they labeled the N400 – this is shown in the dashed line in Fig. 22A. Moreover, strength of this N400 potential is proportional to just how unexpected a particular word might be (Kutas and Hillyard, 1984). The N400 is generally strongest over the central-posterior part of the scalp; this topography is shown in Fig. 22B.

While the N400 was first measured with sentences, it was quickly discovered that the N400 can also be elicited by individual words.[1] For example, the N400 is smaller for words that are used frequently, and is also reduced when words are semantically primed.[2] The semantic priming effect is illustrated in Fig. 22C. In fact, the N400 can even be elicited by non-linguistic stimuli; a priming effect can also be seen for pictures, as shown in Fig. 22D. Results like these have led to the *semantic memory theory* of the N400: This brain response reflects neural processes involved in the activation of meaning from semantic memory. More activation is required, under this theory, to access the meanings of unexpected words. Indeed, you might already have noticed that the timing of the N400 makes sense for this sort of theory: the N400 begins around 200 to 250 milliseconds after the onset of a word (this is perhaps clearest in Fig. 22C) and aligns quite well with the time-course of word recognition that we saw in Chapter 5 (see especially Fig. 18 on page 93).

The *semantic memory theory* receives support from studies showing that the N400 is in fact reduced by words that are entirely unexpected, just so long as they are semantically related to predictable items. To demonstrate this, Kara Federmeier and Marta Kutas (1999) presented participants with sentences like the following:[3]

(1) 'Checkmate.' Rosalind announced with glee.
She was getting to be really good at...

The sentence could be completed by the expected word "chess" or one of two unexpected words: "Monopoly," which belongs to the same semantic

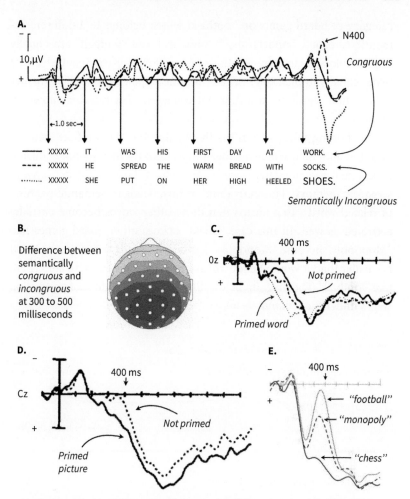

Figure 22. The N400 event-related potential. (A) The first observation of the N400 showed a stark contrast in the brain response to visually incongruous words (dotted line) and semantically incongruous words (dashed line). Note that here, and in all panels in this figure, negative is plotted upwards. (B) The N400 typically has a central-posterior topography. The N400 can be elicited by single words (C) and even meaningful pictures (D), and is reduced when words or pictures are primed. (E) Words that are semantically related to an expected item, even if the word itself is unexpected, lead to reduced N400 responses. This result supports the *semantic memory theory* of the N400.

Sources: A: Kutas and Hillyard (1980); B: Kutas and Federmeier (2011); C: Holcomb (1988); D: Holcomb and McPherson (1994); E: Federmeier and Kutas (1999).

category of board-game, or "football," which belongs to a different semantic category. Importantly, "Monopoly" and "football" are equally implausible ways to complete this sentence for the participants. Despite being equally implausible, words from the same category elicit smaller N400s than words from a different category. This result is shown in Fig. 22E.

This finding follows from the theory that the N400 reflects semantic activation. As a sentence's meaning is built up incrementally during comprehension, predictable words, like "chess," become *pre-activated* in semantic memory. This activation, in turn, leads to semantic priming of related words; that means that those other words become partially activated as well. In this case, "chess" primes other board games like "Monopoly." Because "Monopoly" has been partially activated, the N400 is smaller than it would be for words that have not been activated at all, like "football." The diagram below illustrates the sequence of processing stages under the *semantic memory theory*:

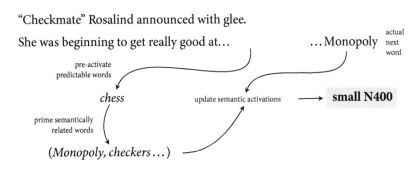

The N400 illustrates how aspects of sentence meaning can affect, or modulate, various stages of processing. But there is an alternative account of the N400 that links it more closely with the computation of sentence meaning itself. The *semantic integration theory* of the N400 holds that this neural response reflects the plausibility of the meaning that results when a new word is combined (integrated) with the previous context. The following diagram sketches the stages of processing that lead to an N400 under the *semantic integration theory*:

"Checkmate" Rosalind announced with glee.
She was beginning to get really good at... ...Monopoly *actual next word*

generate meaning for context → *activate word* → *integrate meaning* → **N400**

Notice how both theories predict that there will be a larger N400 for words that don't match the prior sentence context, but they associate the N400 with different processing stages. The *integration* theory links the N400 with a compositional process of building up the meaning of a sentence, while the *semantic memory theory* links it with the activation of individual word meanings.

In fact, current data indicate that both theories may, in part, be right. To test this, we look to a strategy that we first saw when discussing how to tease apart different aspects of processing during word recognition ("single-trial analysis"; see around page 91). You can think of the degree of match between a word and prior sentence context in two distinct – yet correlated – ways. One way, by now pretty familiar to us, is in terms of whether a word is predictable or not. You can test for a word's predictability using something called a *Cloze task*. A Cloze task is like a "fill-in-the-blank"; you give participants some incomplete sentence and ask them to put in the word that they think fits best. The words that are most commonly filled in are said to have a "high Cloze probability" – they are predictable. But another way to test whether a word matches its context is in terms of its semantic plausibility. A *plausibility judgment* complements the Cloze task: You give participants a full sentence with the target word, and ask them to judge how plausible the meaning of the whole sentence is. Now, you are probabably thinking that these two kinds of measurements are awfully similar to each other; something that is implausible is going to be pretty unpredictable. And you are right! But just because these are similar doesn't mean that they are identical. To see this distinction, consider the following sentence:

Lisa drank a glass of...

<small>what do you predict?</small>

<small>ready?</small>

... guava juice.

I suspect that if you were asked to "fill in the blank" in this case, almost no one would write "guava juice." In other words, the Cloze probability would be 0. But none the less, the situation of someone drinking a glass of guava juice is quite plausible (and delicious).

A large-scale study, comprising efforts from nine different laboratories, took advantage of how predictability and plausibility can be disentangled across individual sentences.[4] Over 300 participants listened to sentences with a noun that did or did not match the context. While the highly predictable words were also highly plausible, words that were not predictable were judged to span a range from high to low plausibility. When they examined the correlation between word predictability (from a Cloze task) and the EEG signal, they observed more positive voltages (that is, less negative) for more predictable words. This is an N400 effect that peaked between 300 and 400 milliseconds after word onset (see the top left of Fig. 23A). Separately, increased sentence plausibility also led to more positive (less negative) voltages, but this effect was slightly later, peaking around 500 milliseconds, and also was measured over a smaller area of the posterior scalp; see the top right of Fig. 23A.

These data indicate that predictability is the principal driver of the N400 effect, which is consistent with the *semantic memory theory* of the N400. But plausibility also seems to have an effect, albeit a smaller one, consistent with the *semantic integration theory*. Notice also that the effect for plausibility is later than that of predictability. This is exactly what you would expect based on the time-course of word recognition that we have already seen: (i) words are pre-activated by context; (ii) lexical access is further modulated when the next word is encountered, 250–400 ms after word onset; (iii) which is followed by the update of the meaning of the whole sentence by around 400–500 ms.

We've established that the N400 ERP component is sensitive, at least in part, to whether or not a word is predictable given its sentence context. This effect reflects the fact that predictable words can be *pre-activated*

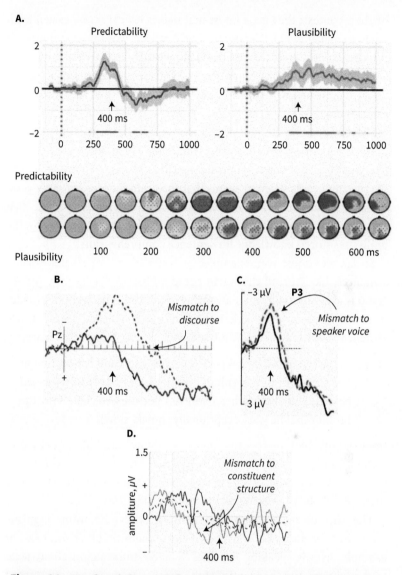

Figure 23. Predictability and the N400. (A) Both predictability and plausibility affect the N400, with the predictability effect appearing earlier and with larger amplitude; remember: more predictable = less negative (smaller N400) = more positive. Different sources of information affect predictability and, consequently, the N400, including (B) discourse, (C) the identity of the speaker, and (D) grammatical structure. (Pay attention to *y*-axis labels!)

Sources: A: Nieuwland et al. (2019); B: van Berkum et al. (2003); C: van Berkum et al. (2008); D: Brennan and Hale (2019).

by their context; they get a boost that makes lexical access easier, leading to a smaller N400. Because the N400 reflects word predictability, we can use it to test different theories of how people make predictions during sentence processing. In particular, I will focus on evidence about the kinds of information that people take advantage of when they make such predictions.

The brain uses quite a lot of different sources of information to make predictions during sentence processing. As we're about to see, it seems like we use just about any information that we can use. But we'll also see puzzling cases where it seems that the brain ignores perfectly good information about what words to expect next. These instances of "bad predictions" may actually point towards some deep principles of how the brain uses stored knowledge for sentence comprehension.

So far, we've just seen examples of N400 effects for single, isolated sentences, like "spread the warm bread with socks" (Fig. 22). But an N400 is also observed when a word is unpredicted based on a broader discourse context, even if it is perfectly well matched to the single sentence that it appears in (Van Berkum et al., 2003). Here's an example:

(2) As agreed upon, Jane was to wake her sister and her brother at five o'clock in the morning. But the sister had already washed herself, and the brother had even got dressed. Jane told the brother that he was exceptionally {quick, slow}.

Importantly, the target words "quick" or "slow" are both equally predictable (and plausible) if you consider just the final sentence alone. But, because a word like "slow" doesn't match the broader context of this little discourse, it elicits an N400, as illustrated in Fig. 23B.

The sort of information that matters can come from linguistic input, like the discourse above, or it can come from other sources. For example, *who* is speaking provides valuable information about *what* they might say, and the identity of the speaker seems also to affect word predictions and the N400. For example, in a 2008 study Jos van Berkum of the Max Planck Institute for Psycholinguistics and colleagues used sentences like the following that were spoken by different kinds of speakers:

(3) *Male or female speaker*
"If only I looked like Britney Spears in her last video."
Child or adult speaker
"Every evening I drink some wine before I go to sleep."

The researchers observed an N400 effect at the target word, underlined in the examples above, when that word didn't match what you would expect based on the voice of the speaker (such as a child's voice saying "I drink some wine"). As you can see in the ERP shown in Fig. 23C, the effect for this mismatch is quantitatively smaller than that observed for mismatches against the discourse (Fig. 23B) or sentence context (Fig. 22A, E). But the difference emerges in the same 300–500 millisecond time-window, and on the same central-posterior areas of the scalp, for a speaker mismatch as for the other kinds of prediction effects. The difference in the strength of the voltage difference may be due to the fact that speaker identity doesn't provide quite as strong a clue to upcoming words as a highly constraining sentence, like the chess sentence used in example (1) on page 120.

In some cases, researchers have debated the degree to which some kinds of information are used during rapid sentence comprehension. One point of debate has been the role of abstract hierarchical structure – the constituency that is diagrammed in Fig. 21. The idea is that such constituency can be complicated to compute rapidly, and it may be that in everyday circumstances language users can make a fair guess at meaning using strategies that rely on simpler information, like each word's syntactic category (so, hearing a $Noun_A$–Verb–$Noun_B$ sequence might be quickly interpreted as "A did something to B").[5] To test this, researchers separately computed the probability of words based on the word sequence that preceded them, or on the constituency structure of the sentences that the words appeared in (Brennan and Hale, 2019). The words themselves came from a chapter of the children's story-book *Alice's Adventures in Wonderland*, which participants listened to while EEG was recorded. The idea here was to give the participants a kind of natural everyday task. As illustrated in Fig. 23D, they observe an increase in negative scalp voltages for unpredictable words, but only when predictability was based on the constituent structure of each sentence; this effect was not seen with predictability based on word-sequence information alone.

The take away message from these last few paragraphs is that the brain makes use of many different information sources during sentence comprehension. Each of these sources of information, from the broader discourse, to who is talking, to the detailed constituent structure of the unfolding sentence, provides a clue as to what words might come next. These clues are combined in order to predict upcoming words, thus helping us rapidly and efficiently understand a sentence.

Just how rapidly are these different sources of information combined and used to predict upcoming input? Several pieces of evidence have narrowed down this important question of time-course. This line of research begins with a puzzle nicely illustrated by a 2016 study from Wing-Yee Chow and her colleagues at the University of Maryland. The experiment uses sentence stimuli that have *semantic role reversals*; what this means is that nouns that are usually the *agent* of a verb are instead given as the *patient*. An example is given below:

(4) *Normal semantic roles*
The restaurant owner forgot which customer$_{patient}$ the waitress$_{agent}$ had <u>served</u> during dinner yesterday.
Reversed semantic roles
The restaurant owner forgot which waitress$_{patient}$ the customer$_{agent}$ had <u>served</u> during dinner yesterday.

Notice that the target verb, *served*, is quite predictable when *waitress* is the agent performing the action to the *customer*. In contrast, *served* is quite unexpected in the second sentence, when the role of patient and agent has been swapped. Despite this clear unpredictability, there is *no N400* when semantic roles are reversed, as in the second example sentence. When the sentences are only slightly modified, the N400 "returns." For example, when the patient noun *customer* is swapped instead with the noun in the main clause (*the restaurant owner* in this example), a clear N400 is observed.

These semantic reversal sentences are interesting because the individual words involved are all semantically related (*serve* is related both to *customer* and to *waitress* etc.). So whether the verb is predictable or not doesn't just depend on the meaning of the other words, but requires the comprehender to process the word, to compute its semantic role, and to use those pieces of information together to predict upcoming input. The idea then is that while context can be rapidly used to make predictions by pre-activating semantically related words (using a wide variety of information sources), more complicated information combinations, like narrowing those predictions down based on semantic role information, may take additional time. When the relevant words are right next to each other, as in the example sentences just above, there has not been enough time for this sequence of predictions to unfold:

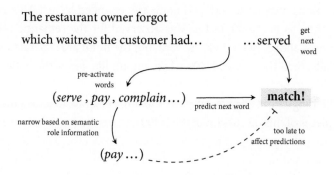

The nature of these limitations on rapid predictions is just beginning to be uncovered, and a more nuanced picture is sure to emerge with further study.[6] Still, there is intriguing data suggesting that comprehenders may just need a few extra tenths of a second in order to incorporate and make predictions based on semantic roles and other argument structure information. Shota Momma proposed a very elegant demonstration of this in his dissertation research, completed at the University of Maryland in 2016. This study takes advantage of a few key properties of Japanese grammar. First, Japanese uses grammatical markers to indicate the subject and the object of a sentence. Second, Japanese has flexible word order, so both the subject (here, the agent) or the object (patient) may appear before or after the verb. Third, arguments may be left unspoken, or "dropped", in grammatical sentences. These three properties allow the creation of semantic role reversal stimulus items:

(5) *Normal semantic role*
    bee$_{agent}$ sting   "The bee stings (something)"
    Reversed semantic role
    bee$_{patient}$ sting   "(Something) stings the bee"

The idea here is that the verb *sting* should be predictable when *bee* is the agent, but not when it is the patient (we don't typically expect bees to be the victims of stinging). Then, the researchers simply varied how much time passes between the presentation of the noun and the verb to probe how long it might take for semantic role information to influence predictions. When the verb and noun are separated by just 0.8 seconds, then there is no N400, just as was observed for the semantic reversal sentences above. But, when the verb and the noun are separated by 1.2 seconds, then there is an N400 when a noun like *bee* is the patient of a verb like *sting*

The diagram in Fig. 24 on page 130 aims to draw together the various pieces of this discussion of the N400 and prediction. There are really just three things happening in this picture. When a word is first accessed (as discussed in Chapter 5), there is additional pre-activation of possible upcoming words. This is shown in the curved solid arrows and the "predicted" items in the bubbles. When the next word is encountered, it leads to what you can think of as an update to the activations of the

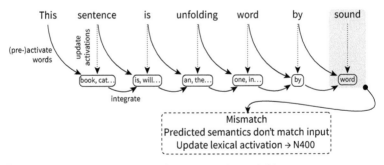

Figure 24. How semantic predictions might unfold. As a sentence unfolds, lexical items are pre-activated, new input updates these activations, and that input is integrated with prior words to guide further predictions. An N400 ERP component may be generated when predictions about lexical items are violated.

mental lexicon. This update is indicated by the dotted vertical arrows. Third, the integration of the current word with the pre-activated words serves to guide further predictions. The key to understanding the apparent complexity of this process is that these processing steps happen again and again as the sentence unfolds. And, as the sentence unfolds, words may become more predictable; this is the case in Fig. 24 where there is a strong prediction at the end of the sentence for the word "word". But, instead, the sentence ends with the word "sound", which was not pre-activated at all. The additional lexical activation needed to access this unexpected word leads to an N400.

The N400 offers evidence that we predict or pre-activate lexical representations while comprehending sentences. We saw in previous chapters that prediction operates at other levels of representations as well – in fact, in some circumstances we may even predict the sensory form of upcoming linguistic input (see pages 58 and 91). Well, listeners may also make predictions about the *syntactic* representations that they expect to encounter during sentence comprehension. Evidence about syntactic expectations comes from a second major ERP component associated with sentence processing, the P600.

The P600 is an ERP component that peaks relatively late, between 600 and 900 milliseconds after encountering a word. It is a positive voltage measured predominantly over the posterior part of the scalp. In contrast to the N400, which seems to be sensitive to semantic mismatches, the P600 is typically seen when encountering a word that syntactically mismatches the context. For example, a 1991 study by Helen Neville and colleagues found a P600 when when participants read ungrammatical sentences like those in example (6) (underlining indicates the phrase from which the P600 was measured):[7]

(6) *Ungrammatical*
 The scientist criticized Lucy's of proof the theorem.
 Grammatical
 The scientist criticized Lucy's proof of the theorem.

Another example comes from a study by Ana Gouvea and colleagues in 2010. They show participants grammatical and ungrammatical sentences like these:

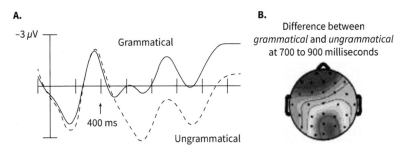

Figure 25. The P600 ERP. (A) Ungrammatical sentences show increased positive response starting about 0.5 second after an ungrammatical word. (B) The P600 is strongest over central posterior parts of the scalp.
Source: Gouvea et al. (2010).

(7) *Ungrammatical*
The patient met the doctor while the nurse with the white dress show the chart during the meeting.
Grammatical
The patient met the doctor while the nurse with the white dress showed the chart during the meeting.

Fig. 25 illustrates the time-course and topography of the P600 effect observed by Gouvea and colleagues.

But the P600 is not only observed when sentences are ungrammatical. It is also found for grammatical but unusual sentence structures. One example of this is from a study by Lee Osterhout and Philip Holcomb (1992), who presented participants with sentences like these:

(8) *More common syntax*
The singer decided to perform the opera.
Less common syntax
The singer allowed to record the song.

The key thing to note here is that the target word, *to*, in the bottom sentence is perfectly grammatical; think about a sentence that continues *The singer allowed to record the song became a big star.* That same sentence could be rewritten as *The singer who was allowed to record the song....* The bottom sentence has what is called a "reduced relative clause" because the words *who was* are optional. Reduced relative clauses like these are less common than sentences like the top one (... *singer decided to...*). While both sentences are still grammatical at the target word, a larger P600 was observed for the sentence with a more unusual structure. Results like this suggest that the P600 signal may reflect expectations (predictions) about likely syntactic structures.

There is significant debate about exactly what is going on in the brain when it generates a P600 ERP.[8] I'm only going to talk about a little part of this interesting line of research that connects back to the broader point in this section about how predictions matter for almost all aspects of how the brain makes sense of sentences. I'm going to discuss an apparent puzzle for the idea that the P600 connects, in some way, to syntactic processing (notice that I'm staying kind of vague about exactly what kind of "syntactic processing" is going on). The puzzle is nicely illustrated in an EEG study by Albert Kim and Lee Osterhout (2005). They presented participants with sentences that began in the following way:

(9) The hearty meal was devoured...   *Semantic match*
      The hearty meal was devouring...   *Semantic mismatch*

Notice that the second sentence is unusual for apparently semantic reasons: Meals don't tend to devour people, people devour meals. Despite this apparent semantic mismatch, the ERP response to sentences like those in (9) is not an N400, as you might expect, but rather a P600.

To make sense of this apparent puzzle, let's go back to *why* an N400 might occur in the usual cases. The *semantic memory theory* holds that the N400 reflects the activation of a lexical item; it is larger for unpredictable words because more activation is required for words that haven't already been pre-activated. The key idea to grab onto here is that the N400 doesn't reflect the prediction match (or mismatch) itself, but rather reflects what the brain does *after* a prediction has failed. So if a prediction is violated, the brain may need to activate other lexical items, and this leads to an N400 signal. But what if the brain does something else when a prediction is violated? Instead of reactivating lexical items, for example, perhaps the brain could change the syntactic analysis of the sentence to accommodate the words that it has already activated.

Something like this *syntactic reanalysis* may be going on for the "semantic" P600 effect described above. A noun phrase like *the hearty meal* pre-activates related lexical items (*eat, devour, enjoy...*), and it also may activate appropriate syntactic representations. Typically, meals are the patient, not the agent, of an action (you serve or eat a meal). So, it could make sense to pre-activate a syntactic structure such as the *passive voice* where the object of the verb, not the subject, is uttered at the beginning of the sentence. To draw out this hypothesis, let's augment the picture from Fig. 24 with additional predictions about syntactic structure. These are shown in Fig 26 by the syntactic diagrams in bubbles at the bottom. These bubbles capture in a very simple way the idea that comprehenders don't just pre-activate words when processing a sentence; they can pre-activate other linguistic representations, including syntactic phrases and dependencies. Now, when a word like *devouring* is encountered, there is a mismatch between the syntactic features of this word – it is an active verb form – and the passive syntactic features that had been pre-activated.[9]

Predictions are dynamic

Before wrapping up the discussion about how the brain makes predictions during sentence comprehension, I want to mention some different kinds of evidence for two key pieces of the dynamic process shown in Figs 24 and 26: making predictions at multiple levels simulatiously, and the interplay between making and updating predictions.

Clues that the brain makes predictions about different kinds of linguistic representations simultaneously comes from an fMRI study led by Alessandro Lopopolo (2017). Participants in this study listened passively to a series of narratives during fMRI scanning. (These naturalistic narratives offer another contrast to the ERP studies above, which relied on carefully constructed "odd" sentences.) To tease apart predictions at different linguistic levels, the researchers considered the narrative stimuli in three different ways. First, they considered

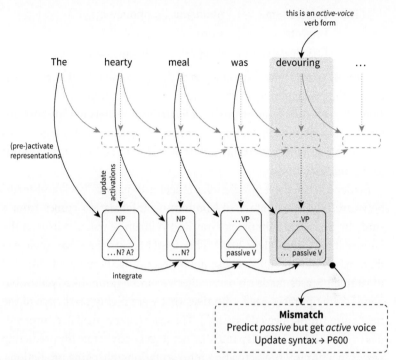

Figure 26. How syntactic predictions might unfold. Predictions unfold at multiple linguistic levels simultaneously, including syntactic structure; here, the word *devouring* violates the syntactic prediction for a passive verb phrase, leading to a P600 ERP response.

the phonemes that make up the narratives; by counting how often phonemes co-occurred, they quantified the probability of the phonemes in each word. Second, they considered the probability of each word by counting how often certain sequences of words appeared together. Lastly, they considered syntactic probability in terms of how often certain syntactic categories appeared in order together. Each of these measures can be written out as a *conditional probability*, shown in (10). You can translate these statements into English along the lines of: "The probability of a given word, call it w, following two previous words, w_1 and w_2, is...."

(10) $\Pr(\text{phoneme}_p \mid \text{phoneme}_{p-1}, \text{phoneme}_{p-2}...)$
$\Pr(\text{word}_w \mid \text{word}_{w-1}, \text{word}_{w-2}...)$
$\Pr(\text{category}_c \mid \text{category}_{c-1}, \text{category}_{c-2}...)$

The fMRI scans show distinct areas of the temporal lobe that are sensitive to the predictability of these different representations. Unexpected syntactic categories, for example, activate both posterior and more anterior areas of the left temporal lobe, while unexpected lexical items activate an area that covers the middle of the temporal lobe, just posterior to the auditory cortex.

Evidence that these predictions reflect a dynamic "back-and-forth" between linguistic context and bottom-up information comes from a study by Anastasia Klimovich-Gray and William Marslen-Wilson at the University of Cambridge (2019). They used MEG to track how such predictions unfold and affect subsequent processing. Participants listened to simple phrases made up of an adjective and a noun, like "yellow banana." Some phrases used adjectives that were highly predictive of the next noun (think "peeled banana"). The researchers calculated three values from their stimuli to tap into distinct stages of dynamically predicting words in a phrase: pre-activating representations, checking predictions, and updating activations. Pre-activation was tested by comparing adjectives that were more or less predictive of the next noun. The step of checking predictions was tested by computing the match between predictions from the adjective, and the first phoneme of the noun (e.g. [b] from "banana"). The first of these values can be computed using variants of conditional probabilities, as discussed in the paragraph just above. The last semantic value is the simple average of all of the word embeddings that make up the phrase (take a look back at the diagram on page 103 for a refresher on word embeddings).

MEG affords the researchers remarkable precision in tracking these processing stages in both space and time; the results are summarized in Fig. 27. They show an effect for pre-activation even before the end of the first word. This effect is in the left frontal lobe, specifically the inferior frontal gyrus. This region has long been closely associated with language processing in the brain (see page 10), and we see here the first clues that it may play an important role in sentence comprehension; we'll see more about this region in Chapter 9. Just a few hundred milliseconds later,

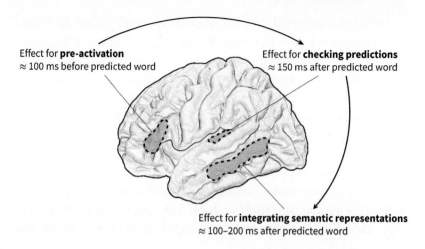

Figure 27. Prediction dynamics. Predictive processing, localized to the frontal lobe, affects sensory predictions and semantic integration to subsequent linguistic input in the temporal lobe.
Source: Adapted from Klimovich-Gray et al. (2019).

the neural data show evidence for those predictions being checked as new information from the second stimulus word – the noun – enters the auditory cortex. Around the same time, activity in the posterior middle temporal gyrus is sensitive to the integration, or semantic combination, of the new noun with the just-heard adjective.

Chapter summary

This chapter introduced the foundations of how the brain makes sense of sentences.

- To understand a sentence you must decode the *constituency*, or structural relationships, as well the *dependencies* between words. Meaning is a *compositional* function of the words and how they are structurally put together.
- To do this efficiently and rapidly, the brain is constantly making – and checking – predictions for what might come next in the sentence.

- When confronted with an unexpected word, additional lexical activation is associated with the N400 ERP component that is observed as a negative voltage potential 300–500 milliseconds after an expectation is violated on central-posterior areas of the scalp.
- When confronted with unexpected syntactic structure, syntactic reanalysis is associated with the P600 ERP component as a positive voltage at around 500–800 milliseconds on posterior areas of the scalp.
- Linguistic predictions are based on a wide variety of clues, including the broader discourse and social context of an utterance. That information is used to shape expectations about multiple levels of linguistic representations, including phonemes, words, and syntactic categories.

One larger take way from this section is that when we study sentence understanding in the brain, we are almost always looking at processing of one or more words in some context. So we have to contend with the predictions that have been made, and consider how the brain might likely deal with some new input given that context. Keeping this firmly in mind, we next turn to the neural operations that go into the building of sentence structure itself.

8

Composing sentences

In the previous chapter, we reviewed how sentences are structured and how the brain makes use of predictions in multiple ways to help make sense of that structure rapidly and efficiently. Prediction is of fundamental importance in characterizing this capacity, but it alone does not *explain* how we understand sentences. Simply put, being able to predict what comes next is not the same as understanding what someone is saying. In order to understand, the brain must put words together to form constituents, and also recognize the longer dependencies that undergird sentence meaning. This chapter takes up how the brain forms constituents, while Chapter 9 turns to dependencies.

A combinatoric network

The earliest research into how the brain puts sentences together contrasted sentences with structure to stimuli without structure, like word lists:[1]

(1) the man on a vacation lost a bag and a wallet　　*Sentence*
　　 on vacation lost then a and bag wallet man then a　*Word list*

These two types of stimuli differ in many ways. In addition to having syntactic structure, the top sentence is also easily interpreted – it has a clear meaning, and that meaning is even quite plausible. Moreover, and I expect you're already thinking this, real sentences are much more predictable than word lists.

And indeed, many studies using stimuli like these show that a wide range of brain regions are more activated by sentences than word lists.

Language and the Brain. Jonathan R. Brennan, Oxford University Press.
© Jonathan R. Brennan (2022). DOI: 10.1093/oso/9780198814757.003.0008

To give one example, Christophe Pallier and colleagues (2011) presented participants stimuli like those shown in Table 5 while recording brain responses using fMRI. These stimuli include twelve-word sentences, word lists, and also intermediate stimuli with phrases comprising two, four, or six words (Table 5). As expected, the researchers observed a broad range of regions in the left superior temporal gyrus and inferior frontal gyrus that show greater activity for sentences that have more phrases (and are more interpretable, plausible, predictable, etc.); a sample of their results is shown in Fig. 28A. These regions each show a steady increase in activation for stimuli with more structure (i.e. larger and larger phrases).

We can think of this broad set of regions as a kind of *combinatoric network* of brain areas that are engaged in some way when understanding sentences. Of course, it is not immediately clear what functions might be reflected in these different regions given the many differences between sentences and word lists. To reduce this ambiguity, at least a

Table 5. **Stimuli from Pallier et al. (2011).** Examples of sentence, phrase, and word list stimuli from Pallier et al. (2011) (adapted from French). The pseudoword items are less predictable than the real word items.

Words per phrase	Stimuli
	REAL WORDS
12	[I believe that you should accept the proposal of your new associate]
6	[the mouse that eats our cheese] [two clients examine this nice couch]
4	[mayor of the city] [he hates this color] [they read their names]
2	[looking ahead] [important task] [who dies] [his dog] [few holes] [they write]
1	thing very tree where of watching copy tensed they states heart plus
	PSEUDOWORDS ("Jabberwocky")
12	[I tosieve that you should begept the tropufal of your tew viroate]
6	[the couse that rits our treeve] [fow plients afomine this kice bloch]
4	[tuyor of the roty] [he futes this dator] [they gead their wames]
2	[troking ahead] [omirpant fran] [who mies] [his gog] [few biles] [they grite]
1	thang very gree where of wurthing napy gunsed they otes blart trus

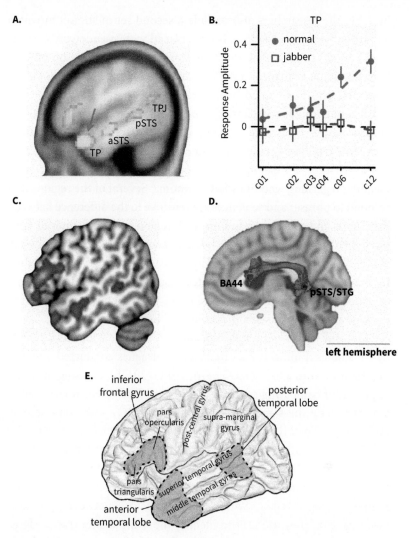

Figure 28. A combinatoric brain network. (A) Regions that show increased activity for larger phrases span the left temporal and inferior frontal lobes. (B) Some of these regions, like the temporal pole, show more activity when sentences contain real words (circles) but not pseudowords (squares). (C) That set of fronto-temporal brain regions are also observed in a meta-analysis of 19 studies. (D) The same study also found that the white-matter pathway directly connecting posterior temporal and inferior frontal regions is aligned with the peaks of brain activity for composition. Panel (E) schematically illustrates the primary foci of the combinatoric brain network.

Sources: A and B: Pallier et al. (2011); C and D: Zaccarella et al. (2017b).

little bit, the researchers also include a second set of stimuli in which all of the content words have been replaced with pseudowords. These pseudoword sentences are illustrated on the bottom rows of Table 5. By retaining grammatical function words, the stimuli still have phrase structure. These are called "jabberwocky" items after the famous non sensical poem in Lewis Carroll's *Through the Looking-Glass*. Without content words, the stimuli become much less predictable. Of course, such stimuli still retain quite a bit of complexity in addition to the syntactic structure; for example, even pseudoword sentences still convey meaning regarding who did what to whom.[2] Several of the regions that respond to phrases and sentences are sensitive to the difference between real words and pseudowords. One such region is the left anterior temporal lobe (or *temporal pole*, TP), whose activation pattern is shown in Fig. 28B. Activity in this region only increases for larger phrases when they contain real words. This may tease out sub-parts of this combinatoric network that are sensitive specifically to predictions, or perhaps to the kind of referential semantics conveyed by real words.

This same fronto-temporal *combinatoric network* is found in a number of studies. Emiliano Zaccarella and colleagues conducted a review of 19 studies with a total of 295 participants (2017b), all using the same approach of comparing sentences with word lists. Their review shows almost exactly the same set of regions (Fig. 28C) with some important nuance. First, their review again highlights the *left inferior frontal gyrus*, or LIFG, as a region that appears to play a significant role in sentence understanding alongside anterior and posterior temporal regions. Second, they augment their review with an analysis of the *structural connectivity* between regions within the combinatoric network using DTI (see page 26). They use DTI to identify the white-matter tracts – large bundles of axons connecting one population of neurons to another – which go from the temporal lobe to the inferior frontal lobe. It has long been known that there is strong structural connectivity between posterior temporal and inferior frontal regions; this anatomical connection is called the *arcuate fasciculus*. This particular analysis shows how that pathway, which is illustrated in Fig. 28D, seems to connect the specific "hot spots" – or centers of neural activity – that are implicated in comprehending sentences.

The *combinatoric network* shown in Fig. 28 is a reasonable starting point for studying how the brain puts sentences together. But to uncover the contribution of each of the individual nodes, and how they relate to each other, we need to turn to a different set of experimental methods. A key limitation of comparing sentences with lists, and with variants like pseudoword "jabberwocky" sentences, is that we don't have a deep understanding of what participants *do* when they are confronted with stimuli of this sort. Do participants try to figure out phrases and meanings even from random lists, or do they give up? In the review study discussed above Zaccarella et al. (2017b), the authors identify interesting differences between studies that constructed word lists in different ways (such as by including or excluding grammatical function words); perhaps participants try to make phrases when they see function words, even if the words are all jumbled. The issue here concerns the *linking hypothesis* connecting the experimental design – the stimulus, task, and so forth – with the kind of neural processes that are required to perform the experiment. We only have a limited understanding of how those links work in the case of word lists, and consequently the conclusions we can make about, say, differences when processing real words or pseudowords are likewise limited.

So, what sort of experimental tools and linking hypotheses can we use to dive deeper into this combinatoric network?

Composing simple phrases

One alternative approach is to study compositional processing in a more minimal way, by focusing, for example, just on what makes a two-word phrase different from two words that do not form a phrase. This approach, which I'll call "simple composition," has fewer variables compared with full sentences, making for a simpler linking hypothesis. That, in turn, helps researchers draw out more specific conclusions about the function of the individual pieces of the broader combinatoric network.

The basic idea is illustrated by the experimental design in (2), below, which was developed by Douglas Bemis and Liina Pylkkänen at New York University (2011).

(2)

	Composition task		List task	
Two words	red boat ⛵		cup boat ⛵	
One word	xkq boat ⛵		xkq boat ⛵	

Two factors are manipulated here in order to isolate, in a minimal way, the processes that combine two words into a single phrase. The first factor is whether the stimulus contains two words or just one. This is shown across the rows in (2). The two columns differ in terms of the task that the participant is asked to do; both are variants of a picture-matching task. For the left-hand column, the task is to answer whether the picture matches the full, composed meaning of the words. So, a "red boat" is a match, and just the word "boat," but "blue boat" would not be a match. In the right-hand column, the task is to answer whether the picture matches *any* of the words. This "list task" does not require compositional processing. Note also that none of these conditions are particularly predictable: a color word like "red" does not raise (or lower) your expectations for the word "boat." So, the idea here is to look for brain activity that increases when composing two words together, compared to any of the other three conditions.

Just such a response pattern was observed using MEG in two brain areas: first, in the *left anterior temporal lobe*, or LATL, at around 250 milliseconds after the onset of the second word, followed by activity at around 400 milliseconds at the anterior tip of the frontal lobe, a region called the *ventro-medial prefrontal cortex*, or VMPFC.[3] These findings are shown in Fig. 29A. Taking advantage of the fine temporal resolution offered by MEG, the researchers reasoned that the LATL activation reflects an initial stage of composition, while the later VMPFC activation reflects a later, perhaps semantic, evaluation based on the composed phrase. Interestingly, there was no indication of activity in the LIFG, which was so prominent in the comparisons of sentences and lists, in this more minimal instance of composition.[4]

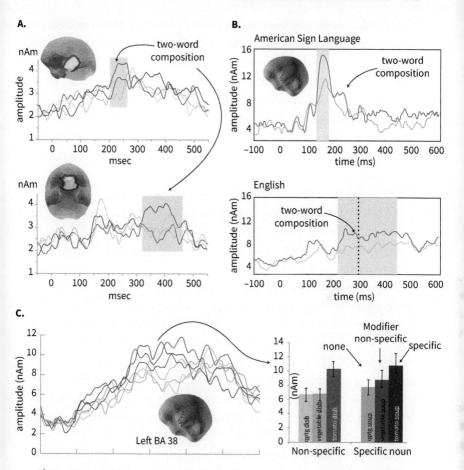

Figure 29. Simple composition and the anterior temporal lobe. (A) Reading simple two-word phrases leads to more left anterior temporal activation, compared to single words (see page 144 for stimuli). (B) This "simple composition" effect in the LATL is also observed for producing phrases in multiple languages. (C) The LATL effect for composition is not strictly syntactic, as it shows a sensitivity to the conceptual specificity of the terms that are being composed; the interaction between semantic and syntactic factors is shown on the right-hand side.

Sources: A: Bemis and Pylkkänen (2011); B: Blanco-Elorrieta et al. (2018); C: Zhang and Pylkkänen (2015).

A number of studies have used variants of the "simple composition" experimental design both to test the generalizability of these findings and to better understand the nature of the compositional processes carried out in these regions. For example, the LATL and VMPFC also show increased activation for phrase composition when participants produce phrases, not just comprehend them. And this effect has been further observed in speakers of different languages, including languages as diverse as English and American Sign Language.[5] Fig. 29B shows MEG results from the LATL from participants who were producing either American Sign Language (top) or English (bottom). Regardless of language, the LATL showed increased activity for phrases, compared to lists. Be sure to note that the timing of the activity between panels A and B of this figure are a bit different because participants are *reading* words for Fig. 29A, but *producing* them for Fig. 29B.

One of the issues I've highlighted repeatedly in this chapter is how sentence understanding depends on many interdependent processes. This complexity still holds even for the relatively minimal phrases, just an adjective and a noun, used in these studies. For example, we saw at the beginning of the chapter that building sentences requires not just identifying dependencies and phrase structure, but also combining the meanings of words together according to those structures (take a look back at Fig. 21 on page 117). In other words, syntactic structure-building is always correlated with semantic combination.

How can we tell whether the activation in the LATL for phrase composition reflects syntactic structure-building or semantic composition? Above, we considered the tentative idea that the earlier (around 250 milliseconds) activity in the LATL might reflect structure-building, while later activity in the frontal lobe might be more semantic. But we've actually already seen some evidence that the LATL itself could play more of a semantic role in sentence understanding. Those clues came from the discussion of a *semantic hub* for the representation of word meaning, which we saw back in Chapter 6 (see Fig. 20 on page 105). That discussion looked at evidence from semantic dementia and other sources that the anterior temporal lobe appears to play an important role in how the brain represents specific concepts.

Let's consider, then, the semantic consequences of composing a phrase: The meaning of a phrase like "red boat" seems, at least intuitively, to refer to a more specific kind of concept than the single word "boat." Could it be, then, that the LATL activity observed for phrases reflects this more specific kind of conceptual meaning, rather than a structure-building process? Linming Zhang and Liina Pylkkänen (2015) aimed to answer this question. They did so by augmenting the "simple composition" experiment with another factor: whether the modifier or noun, individually, denoted a more or less specific concept. An example of their stimuli is shown in (3).

(3)

Noun:	Modifier:		
	None	*Non-specific*	*Specific*
Non-specific	xpt dish	vegetable dish	tomato dish
Specific	xpt soup	vegetable soup	tomato soup

Here, "soup" and "tomato" each refer to more specific concepts than their counterparts, "dish" or "vegetable." The evidence from semantic dementia that we discussed back in Chapter 6 suggested the LATL would be more engaged by the specific concepts than by the non-specific concepts. As "tomato soup" is even more specific than "tomato" or "soup," that same theory predicts a simple increase in activity for phrases compared to words.

But, the researchers observe a more complex pattern of results than what would be predicted if the LATL solely reflected conceptual specificity. This complex pattern is illustrated in Fig. 29C. What they found was that neither conceptual specificity nor syntactic structure-building alone capture the pattern of results. Rather, these factors appear to interact: There is increased activity for structure-building, but only when the words that are being composed have meanings that are relatively specific. For example: "tomato dish" leads to increased LATL activity relative to just "dish," but a phrase like "vegetable dish," with a less specific modifier, does not.

While the pattern of results is a bit complicated, the bigger take away supported by the data at hand is actually rather elegant. And it shows how observations from very different methods addressing different questions can begin to converge into a consistent picture. The picture seems to be this: The LATL is involved in a process that combines the conceptual features of words. This theory – which is still being developed – makes the interesting prediction that phrases that don't derive a more specific concept will not involve additional LATL activity. This prediction appears to be correct. For example, phrases that involve syntactic structure but don't alter conceptual features, like the number phrase "two boats," don't seem to affect LATL activity (Blanco-Elorrieta and Pylkkänen, 2016).

Still, challenges remain. One chief challenge comes from looking again at patients with damage to anterior temporal regions. These patients, who do show semantic dementia, do not appear to have any systematic difficulties understanding compositional meaning.[6] An open line of research concerns how neural systems may change, or compensate, in response to damage. One way forward is to consider carefully how the network of brain regions involved in word – and sentence – understanding might work together (recall Fig. 28), and maybe even redundantly, for effective understanding.[7]

A second major challenge concerns the role of different sub-regions within the LATL. Notice how the LATL area outlined in Fig. 28E spans the anterior portions of the superior, middle, and inferior frontal gyrus. This region also includes the temporal pole where the gyri converge at the anterior tip of the temporal lobe. This large part of the temporal lobe contains many neural circuits that are involved in many different cognitive processes. Indeed, the maps of the combinatoric network in Fig. 28A and C show at least two separate "hot spots" within this larger LATL area. But current research does not neatly identify these distinct hot spots with different syntactic or semantic processes. A chief goal of ongoing research is to better understand the roles of different parts of this larger temporal area.

We turn next to the roles of other nodes in this broader combinatoric network.

Syntax and the posterior temporal lobe

There are quite a number of questions lingering from the discussion of composition so far: (i) What is going on with the LIFG? (ii) What about that other temporal region, the LPTL? And (iii) Do these simple composition results generalize to "real-world" language? For the first question, I'm going to ask you to be patient. We'll get back to the LIFG when we turn our attention to linguistic dependencies in the next chapter. Before that, we turn to the left posterior temporal lobe and also generalizability; these two questions, it turns out, can be addressed together.

A clue as to what role the LPTL plays in sentence understanding comes from an innovative study by Steven Frankland and Joshua Greene (2015). They use *multi-voxel pattern analysis* (MVPA), which was first introduced on page 67 in Chapter 4. Recall, this is a way of analyzing fMRI data that can help to tell whether different kinds of linguistic representations "matter" to a particular brain region, without committing to whether the brain response is supposed to be larger or smaller, or even whether all voxels in the area should respond in the same way. Frankland and Greene focused on the linguistic representation of argument structure. They presented participants with sentences with different argument structures:

(4)

	Role for "ball"	
	Grammatical	Thematic
The truck hit the ball	*Object*	*Patient*
The ball was hit by the truck	*Subject*	*Patient*
The ball hit the truck	*Subject*	*Agent*
The truck was hit by the ball	*Object*	*Agent*

These stimuli vary two factors: the *grammatical role* of a word (whether it is a subject, object, etc.), and also whether it is the *agent* or *patient* of the event denoted by the verb. This second factor is the word's *thematic role*. These factors are separable from each other, and also from individual words (both "ball" and "truck" fill all possible roles in the example shown).

Using MVPA, the researchers test for small sets of voxels whose activation pattern can indicate, or predict, whether a word was presented as the agent or patient of a sentence. They find just such a region in the LPTL. Moreover, in a follow-up experiment, this method revealed two distinct adjacent parts of this LPTL region; activation in one part reliably indicated which word was the agent of a stimulus, while the second reliably indicated which word was the patient (Fig. 30B).

To understand this pattern of results, think about a folder on your computer. Maybe this folder contains all of your digital photos. The contents of the folder will change over time – perhaps quite often if you take a lot of pictures. Despite this change of contents, you always know where to look on your computer for your latest photo. These brain regions may do the same sort of job as a computer folder. But instead of answering "What's my latest photo?" these regions answer a different question, like "Who is performing the action?" or "What is the action being performed on?" In computer science terms, these brain regions act like a *register* – they reliably store a certain kind of information even as that information changes over time.

What about the timing of the LPTL as it supports sentence comprehension? These fMRI data don't provide the same sort of timing information that we saw with MEG, on page 144, showing LATL activation for composition at around 200–300 milliseconds. William Matchin, Ellen Lau, and their colleagues (2019) combined the spatial precision of fMRI with the temporal precision of MEG to address this gap. In their study, participants performed the same task during fMRI recording and, separately, during MEG recording. Connecting with the work we've already discussed, they presented participants with stimuli composed of simple phrases, or whole sentences, and also varied whether the stimuli had normal content-words or pseudowords with less semantic content, shown in example (5). (You'll recall from page 143 some of the complications that might arise when participants encounter such "jabberwocky" stimuli; I'm not going to discuss those results here).

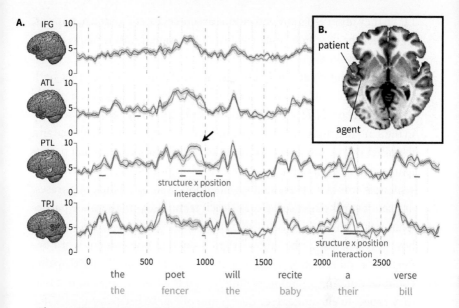

Figure 30. Syntax in the posterior temporal lobe. (A) Combining fMRI and MEG methods reveals an increase in left posterior temporal activity for sentences, compared to phrases, at the beginning of the sentence (arrow). This pattern is consistent with composing syntactic structure predictively. (B) Activity in distinct parts of the left posterior temporal lobe appears to act like "registers" that encode the thematic role of a noun, such as whether it is the agent or patient of a verb.
Sources: A: Matchin et al. (2018); B: Frankland and Greene (2015).

(5)   *Content words*
    Phrase   [the fencer] [the baby] [their bill]
    Sentence   [the poet will recite a verse]
                    Pseudowords
    Phrase   [the tevill] [the sawl] [their pand]
    Sentence   [the tevill will sawl a pand]

The fMRI data from their study revealed several now-familiar nodes of activation along the left temporal and frontal lobes. These regions are shown on the left side of Fig. 30A. The MEG data furnished a millisecond-by-millisecond record of activation in these regions. In the

LPTL they observed more activation for sentences, compared to phrases, right after the very first content-word of the stimuli; this is marked with a black arrow in Fig. 30A.

Stop right here and look carefully at this figure. You'll see that both of the stimuli illustrated at the bottom begin with a simple noun phrase (e.g. *the poet* or *the fencer*). Well, how would a participant know so early in the sentence that one of these stimulus items will be a full sentence and the other will be just a set of separate phrases? The answer is that participants viewed each of the types of stimuli in short *blocks*; so a set of full sentences would be presented together, followed by a block of phrases, or pseudoword sentences, etc. For each stimulus item, then, the participant has a idea of what sort of structure it might have. This explains how participants might "know" right away that one content-word belongs to a full sentence, while the other doesn't. Basically, participants could be building a kind of scaffolding for the rest of the sentence that they expect to encounter (see the bottom of Fig. 26 back on page 135 for an illustration of this), and this building process seems to evoke LPTL activity.

There are other interesting things happening in these results that I don't want to dwell on here (including phrasal effects in the more posterior *temporal-parietal junction* (TPJ) area as well as semantic effects in the LATL that line up with our earlier discussion of that region). The key take away is that processing more complex phrase structures, such as is found in full sentences, leads to more activity in the LPTL, and the timing of this activity is consistent with the kind of syntactic predictions that we've already seen earlier in this chapter.

More evidence linking the left temporal lobe with composition, and linking the LPTL specifically with composing argument structure, comes from efforts to probe linguistic composition in a way that generalizes to more everyday uses of language. Rather than read or listen to a series of phrases or single sentences, each unrelated to the next, participants in these more *naturalistic* studies simply read a narrative text or listen to an audiobook story. The idea is that these sorts of stimuli engage the same sorts of language processing that we use on an everyday basis, such as when we listen to an audiobook while taking the bus.[8]

Table 6. Quantifying syntactic features. Wehbe et al. (2014) quantified a range of linguistic features in a popular novel. The features span different levels of language processing. Shading indicates the presence, and strength, of each feature word by word.

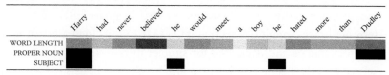

Leila Wehbe and colleagues at Carnegie Mellon University (2014) took this approach in an fMRI experiment where participants read a chapter of *Harry Potter and the Sorcerer's Stone* while brain activity was measured. They used the MVPA data analysis technique, mentioned above on page 149. This technique, you'll recall, can indicate if a brain region is sensitive to certain kinds of linguistic information. Because participants read from a rich natural text, the researchers annotated many different kinds of linguistic information. Table 6 gives a somewhat simplified example. Here, each row corresponds to a linguistic feature examined in this study; gray shading indicates the presence or absence of that feature at each word in the story. The idea is to build a statistical model that uses the fMRI data to predict what the value of each feature ought to be. Clusters of fMRI voxels that do this well are said to "encode" that particular feature. As you can see from Table 6, the researchers encoded different kinds of linguistic features; WORD LENGTH captures something about the visual stimulus itself, while SUBJECT is a syntactic feature that relates to argument structure. And indeed, syntactic features were reliably encoded by voxels in the LPTL.

The syntactic features used in this study address a similar kind of question to the study of argument structure processing that we saw on page 149. Other naturalistic studies use syntactic features that align with whether the sentence structure is more or less complex, along the same lines as the simple composition studies discussed earlier in this chapter. In one such, participants listened to an audiobook chapter from *Alice's Adventures in Wonderland* during fMRI scanning (Brennan et al., 2016). The researchers annotated the stimulus with phrase structures, similar to what we saw at the beginning of the chapter in Fig. 21. Then, they counted how many phrases were completed by each word in the stimulus. You can get a sense of this by going back to Fig. 21: The word *table* completes four phrases (including the sentence itself), the word *cut* completes three phrases, and *apple* completes two phrases. With these measurements, they tested whether the fMRI signals followed the pattern of phrase complexity such that the fMRI signal increased for more complex phases and decreased for simpler phrase structures. They observed exactly this pattern in three brain regions: the LPTL, the LATL, and the LIFG.

These sorts of studies, which use more natural "everyday" stimuli like narrative texts or audiobooks, are powerful tools to test how well observations about the brain generalize beyond a highly controlled laboratory experiment. The results seem to show reasonable agreement with more controlled studies. To review: The results are consistent with a LPTL region that is involved with processing phrase and argument structure, as well as a LATL region that is involved in the composition of semantic representations.

But, as you may have already noticed, these naturalistic studies require the researcher to commit to a very precise account, or model, of what they think the brain is doing when processing the text. We saw, for example, in Table 6 that the researchers converted different linguistic features into precise numbers. As in any ongoing research, these models are unlikely to be fully correct; I expect future work to improve on the sorts of models that are used to specify the linguistic features being processed during these stimuli. Notice the great opportunity here! These models depend on careful reasoning about the kinds of linguistic structures that the brain is using, and so this kind of work really highlights the benefits when careful linguistic theory is connected with neural methods.

Other ongoing work also aims to better understand distinctions between different aspects of the posterior temporal lobe. As with the LATL, the LPTL constitutes a very large area of the cortex. The combinatoric network illustrated in Fig. 28 actually shows two "hot spots": one spanning the posterior superior temporal gyrus, and a second, more posterior one that borders on the parietal lobe (the so-called *temporal-parietal junction*, or TPJ). We further saw in this section evidence that different parts of the LPTL may encode different aspects of argument structure. But teasing apart the distinct roles of these finer sub-division is only just beginning.

Chapter summary

In this chapter we saw several different approaches to uncovering how the brain identifies constituents during language comprehension – how it builds sentence structure.

- Sentence processing involves a network of interacting regions that span the *anterior* and *posterior temporal lobe* (LATL, LPTL) of the left hemisphere, as well as the *left inferior frontal gyrus* (LIFG).
- MEG studies reveal increased activation in the LATL even for very simple phrases, not just complex sentences, within just 200–300 milliseconds of encountering a word. This region's function appears to be sensitive both to *constituency* and to the *conceptual specificity* of a phrase.
- The LPTL is also involved in processing constituency and *argument structure*. There is debate about its specific function, but one theory connects a part of this region with building syntactic structure predictively.

We've seen repeatedly that the frontal lobe, not just the temporal lobe, is also involved in sentence processing. You've been very patient as I asked you to wait ("We'll get to it later"). Well, we can now turn to this question, which has been a very vexing one in neurolinguistics for a long time, of what role the LIFG of the frontal lobe plays in sentence understanding.

9
Building dependencies

Travel back in time with me, for a moment, in order to consider the "classical model" of language in the brain that was presented all the way back in Chapter 1 (page 12). That model, which was based on the groundbreaking deficit/lesion work by people like Broca and Wernicke, posited two primary neural centers for language: a region in the left posterior temporal lobe associated with *comprehending words* and a region on the left inferior frontal gyrus associated with *producing words*. First, just take a moment to appreciate how remarkable it is that, over 150 years later, these two areas of the brain are still understood to be central to the brain bases of language. But now, let's acknowledge that the functional role of these areas has been much refined.

Focusing on the LIFG, that classical model divided language into two processes: comprehension on the one hand, and production on the other. More modern understanding, as we've seen throughout this book, has been grounded in dividing language into separate levels of representation like phonemes, lexical items, phrases, and sentences... each of which is invoked in different ways during both comprehension and production. And indeed it has been recognized for over 50 years that deficits in speech production, like non-fluent "Broca's" aphasia, can't solely be understood in terms of production. "Broca's" aphasia has also been linked to comprehension deficits, specifically difficulties with more complex aspects of syntax.

Dependencies, prediction, and memory

Cases of syntactic deficits in non-fluent aphasia were brought to the fore in a series of articles by the neurologist Norman Geschwind (1970;

1972) in the 1960s and 1970s. He observed, in particular, that patients with non-fluent aphasia had difficulty understanding complex sentences when the subject and object of the main verb were semantically interchangeable. Such participants have no difficulty, for example, choosing the correct picture to match the sentence like that shown in (1a). But they do show difficulty for sentences that *seem* equally simple, such as in (1b).

(1)   a. The deer was chased by the lion

b. The tiger was chased by the lion

What sets these two sentences apart? They differ in how the word meanings – the animals used – provide clues about the grammatical and thematic roles for the sentence. In (1a), *lion* is easily understood as the subject – and agent – of the *chase* action; lions naturally chase deer! But, in (1b), there is no such easy clue as to who is doing the chasing. The sentence in (1b) is *semantically reversible*: Both of the nouns could be either the subject (agent) or object (patient) of the verb. The other thing to note about both of these sentences is that they are in the passive voice. In this sort of sentence, the subject and object of the verb are not in their usual position. Instead, there is a *long-distance dependency* that indicates that the grammatical object, which usually appears after the verb in English, is in fact the noun that appears at the beginning of the sentence:

(2)   The tiger was chased ____ by the lion.

So, it seems that patients with non-fluent aphasia have a rather interesting kind of deficit in syntactic comprehension as well: They have difficulty resolving long-distance dependencies without the help of extra semantic clues.[1]

But wait! These early observations linked aphasia symptoms (non-fluent or "Broca's" aphasia) with a syntactic comprehension deficit. They don't directly link a particular brain region, like the LIFG, with such a deficit. Quite a number of studies have sought to make exactly such a link, including case studies of single individuals and large-scale group studies of aphasia patients. One of the challenges raised by those efforts is how to integrate findings from different participants with distinct, but perhaps overlapping, deficit and lesion patterns.[2] Using advanced statistical techniques to correlate patterns of lesion overlap with shared symptoms is one way to address this challenge, as in a study of over 70 stroke patients by Nina Dronkers and colleagues in 2004. Another strategy is to focus on a patient group with somewhat more homogeneous patterns of brain damage and deficits. Just such patterns can be found in neurodegenerative disorders that affect language-related brain regions in a systematic way. These are *Primary Progressive Aphasias* (PPA).[3] In fact, we've already seen an example of one of these before: *semantic dementia* from Chapter 6. Whereas semantic dementia primarily affects the anterior temporal lobes, other examples of PPA lead to selective neural degeneration in posterior temporal regions and frontal regions.

Serena Amici, Maria Luisa Gorno-Tempini of the University of California, San Francisco, and their colleagues conducted a relatively large-scale study in 2007 of syntactic comprehension in 58 patients with variants of PPA.[4] In addition to the collection of MRI data indicating the location of brain damage, all patients completed a comprehensive battery of language tests. These tests included picture-matching tasks, similar to the ones shown in example (1b), that contain long-distance dependencies and, importantly, have arguments that are semantically reversible.

The effects of neurodegeneration on brain structure are different than those for stroke-induced lesions. To measure progressive neural damage, these researchers used something called *voxel-based morphometry* (VBM). This approach measures the thickness of the cortical gray matter across different areas of the brain; see the inset of Fig 31A for an illustration. The key idea is that neural degeneration reduces the amount of gray matter – cell bodies – in a particular region, and this degeneration affects cognition. Indeed, difficulties with the most complex semantically reversible sentences were found to correlate with just such reduced gray matter in the inferior frontal gyrus, including the pars triangularis; this

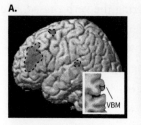 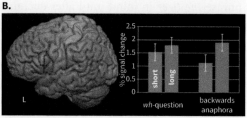

Figure 31. Dependency processing in the LIFG. (A) Primary progressive aphasia patients with inferior frontal neurodegeneration show decreased understanding of sentences with long-distance dependencies. (B) LIFG activation does not solely reflect certain specific long-distance dependencies, like those found in *wh*-questions, but instead seems to correlate with more general predictions that make demands on working memory.
Sources: A: Amici et al. (2007); B: Matchin et al. (2014).

result is shown in Fig. 31A. In other words, patients with less inferior frontal gray matter did a worse job interpreting semantically complex sentences.

These are among the many deficit/lesion studies that compellingly link the LIFG with the processing of long-distance dependencies. Many different theories have been proposed to account for this link; these theories differ in the exact function that is ascribed to the LIFG.[5] One major dividing line is between theories that ascribe a function that is just for language, or *domain-specific*, and theories that link this LIFG processing with a broader *domain-general* function. An example of a language-specific theory is the *trace-deletion hypothesis* developed by Yosef Grodzinsky (Grodzinsky, 2000; Grodzinsky and Santi, 2008). Here, "trace" refers to the part of a long-distance dependency where a word is interpreted – this is the gap that is underlined in the example sentence in (2). The idea is that damage to the LIFG might impair how individuals represent the structure of the long-distance dependency, thus causing poor comprehension for such sentences. An alternative, more domain-general hypothesis is that the LIFG is involved in maintaining a representation in working memory for subsequent processing.[6] Indeed, it's quite reasonable to think that many long-distance dependencies do require extra working-memory resources. For example, to understand the passive sentence illustrated in (2), a listener must keep the noun phrase *the tiger* in mind until after encountering the verb *chased*, like this:

(3) The tiger was chased by...

The working memory demand schematized here is closely related to the kind of predictive processing discussed in the last two chapters; there is indeed a prediction for a transitive verb – a verb that takes an object like *tiger* – later in the sentence.[7] This sort of demand on working memory is actually pretty common for long-distance dependencies. Here is another example of such a dependency that comes from how questions are formed in English:

(4) a. Which song did the band play ___ at the concert?

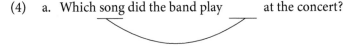

b. Which song did the band play at the concert?

This is called an English *wh*-question because of how the syntax is similar for questions starting with words like *which, what, why,* etc.

Now, looking at the two diagrams in example (4), you might think: "Those look awfully similar to each other; aren't they showing the same thing?" The answer is: not quite, and the difference is actually really important for teasing out what the LIFG might be doing. The diagram in (4a) shows part of the *grammar* of this sentence; this is the mental representation of its structure. In contrast, the diagram in (4b) shows part of the *parsing* process that the brain goes through as it builds this structure. This is a concrete example of the *levels of description* that were introduced way back on page 3 in Chapter 1. The *grammar* of a sentence is part of a *computational* description of the problem the brain is solving, while the *parser* is an example of an *algorithmic* solution to that problem. We must keep this important distinction in mind when we consider the *linking hypotheses* between grammar and brain signals from the LIFG.

In fact, the same sort of parsing processes – and working memory resources – are used even for different kinds of syntactic representations. The sentence in (5) uses a pronoun and a name; both refer to the same individual. There is no "trace" in the syntax here; this is a different grammatical structure than what we saw for the *wh*-question in (4).[8] But, because the pronoun occurs first, when you read this sentence, you must hold it in working memory until you find the noun that "fills out" what it is referring to, as diagrammed in (6).

(5) Because she extinguished the flames, the firefighter saved them.

(6) Because she extinguished the flames, the firefighter saved them.
add to working memory / hold in memory / integrate with noun

Let me add a quick bit of terminology here: the term *anaphora* refers to words, like pronouns, that have their meanings filled in by context. When the pronoun occurs before the word or phrase that describes its meaning, it is called *backwards anaphora*. Let's put the pieces from the last few paragraphs together. Under a *domain-specific* theory of the LIFG like the trace-deletion hypothesis, this region is specifically involved the grammatical representation of certain long-distance dependencies as found in the *wh*-question in (6) but not by the use of backwards anaphora in (5). However, under a *domain-general* theory that links the LIFG to more general working memory demands, both *wh*-questions and backwards anaphora should be supported by processing in the LIFG: Even though they involve different syntactic structures, they make similar demands on working memory.

William Matchin, Jon Sprouse, and Gregory Hickok (2014) at the University of California, Irvine, aimed to tease apart these hypotheses for the LIFG using fMRI. Participants read sentences with *wh*-questions and backwards anaphora that varied in length. The researchers reasoned that longer dependencies should place additional demands on working memory. Just one brain region showed an effect for increased working memory demands, the pars triangularis of the LIFG. As shown in Fig.

31B, this effect was observed both for *wh*-questions and for backwards anaphora. This result supports the idea that at least some sub-parts of the LIFG, like the pars triangularis, are involved in the domain-general working memory demands of complex sentences.

We've just seen an argument linking (parts of) the LIFG with domain-general working memory, not domain-specific linguistic representations. But the stimuli for this experiment were entirely linguistic; it'd be nice to test whether non-language working memory demands also involve the same parts of the LIFG that are seen for long-distance dependences. Actually, just such evidence is available from a familiar source. The 2007 study by Serena Amici and colleagues that investigated neurodegeneration (PPA) and syntax, discussed on page 158 above, also included an examination of non-linguistic working memory. They used a common measure of general working memory: Participants were asked to repeat a list of numbers in reverse order; the list progressively increased in length. Participants with poorer working memory under this measure had decreased cortical volumn in the same part of the LIFG – including the pars triangularis – that was already linked with processing long-distance dependencies.

A natural extension of this line of research is to look at other kinds of linguistic dependencies, such as *agreement* (see Fig. 21). Indeed, both fMRI and aphasia research, especially with agrammaticism (see page 95), implicates the LIFG in agreement processing.[9] But, as always, details matter, and aspects of both neural localization and patterns of grammatical deficit point to both overlap and differences in the neural foundations for agreement as compared to other long-distance dependencies.

Where, when, and what next

These last few chapters together have addressed the remarkable question of how the brain composes words together to make new meanings. The infinitude of expression made possible by human grammar is what sets it apart from any other communication system in the animal kingdom. We've seen that great progress is being made in understanding "where" and "when" the brain builds compositional meaning. Research has mostly focused on three areas of the brain (you may want to refer back to Fig. 28E on page 141):

> **LATL** a left anterior temporal area that appears to be related to composing complex concepts at around 200–300 milliseconds after encountering a word (Fig. 29 on page 145);
> **LPTL** a posterior temporal area that is implicated in predictively computing argument structure (Fig. 30 on page 151);
> **LIFG** the left inferior frontal gyrus that is associated with processing complex sentences, perhaps due to increased working memory demands (Fig. 31 on page 159).

Ongoing work is digging deeper into each of these regions to better delineate the specific neural part (or parts) in each broader area, and to better characterize the syntactic, semantic, and parsing functions associated with each neural circuit.

These answers to the "where" and "when" questions about sentence composition are the first step towards understanding *how* the brain composes sentences. But a complete answer will need more than just the map we've seen so far here. Consider the instruction manual for a new appliance. The first page likely contains a labeled diagram of all the buttons, switches, and components of the device. With this diagram, you can start to explore the rest of the manual to find out how the thing actually functions.

We don't have a manual for the brain, of course, but new efforts have begun to probe such "how" questions. One direction of progress builds on the computational models, discussed briefly in the previous chapter, that aim to quantify more precisely the sorts of computations that specific nodes in the composition network must be making.[10] Another direction of research builds on insights into *neural oscillations*. We were introduced to oscillations back in Chapter 3, where we saw how they might play an crucial role in mapping from continuous acoustic information to categorical speech representations. Now a growing number of studies suggest that oscillations might also play an important role in mapping from strings of words to compositional phrases.[11]

So, consider the resources in these last few chapters as a jumping-off point into this exciting research area.

Chapter summary

This chapter focused specifically on how the brain makes sense of dependencies between words; together with constituents (Chapter 8), this constitutes the "hidden" structure that our brains must uncover to understand what sentences mean.

- Evidence from aphasia demonstrates that the *left inferior frontal gyrus* (LIFG) is involved in processing particularly complex sentences, including those that involve *long-distance dependencies.*
- There are many theories for what specific kind of processing the LIFG could be doing; some theories hold that the LIFG is important for *domain-specific* linguistic representations, others hold that it plays a more *domain-general* role in maintaining things in *working memory.*
- While the question is not yet settled, current evidence from fMRI and the study of *primary progressive aphasia* seems to support the hypothesis that at least some parts of the LIFG play a more domain-general role

This remains a very active research area. As with research into the LATL and LPTL discussed in the previous chapter, one key issue is whether different sub-parts of the LIFG perform different functions. Studies in this area are becoming more sophisticated in teasing apart sub-areas like the pars triangularis and pars opercularis. But it is not clear yet whether these sub-parts are at the right "level" of brain area to link with a specific linguistic function, or whether (and I think this is more likely), even smaller components must be identified. Indeed, careful fMRI research using the "simple composition" protocol has isolated just one part of the pars opercularis (Zaccarella and Friederici, 2015); could some parts of the LIFG be involved in domain-specific phrase-structure representations while others are involved in domain-general working memory?

Another active area of research concerns how the LIFG interacts with other parts of the combinatoric network to carry out domain-specific processing, as shown back in Fig. 28D on page 141. One intriguing hypothesis, developed by Angela Friederici (2017), suggests that the unique compositionality of language emerges from increased structural connectivity between posterior temporal regions and these inferior frontal regions.

10

Wrapping up

In many ways, a book like this one *is* the sum of its parts. My goal has been to give an overview of how neurolinguistics is studied, and what has been learned from this work. We tried to reach this goal this by examining in greater and lesser detail some of the individual research studies and data points that, collectively, contribute to the broader science of understanding how language is instantiated in the human brain. These are the parts. To take stock of how they sum together, this final chapter first checks in to see how we well we achieved the goals that were set out at the beginning of the book. We then look at some of the bigger lessons that have been learned about the brain bases of language, with a focus on where the field may be heading next.

Where are we (in terms of this book)?

There are five broad aims for for this book. Let's check in with each of them.

1. *Introduce the tools of neurolinguistics*

Chapter 2 presented a jam-packed overview of the methodologies used by scientists to study the brain bases of language. The key take aways from these tools are summarized in Figs. 7–9, and especially the "cheat sheet" in Table 3 on page 41. But let's be honest. It's nearly impossible to make sense of the importance of fMRI, the value of MEG, or the unique insights from a TMS study without seeing how these tools are used in practice. It is only when we saw the methods actually put to use to study how the brain processes speech, words, and sentences that we began to engage with the essential issue that different tools are most suitable for answering different kinds of questions.

This match of question with tool deserves reflection. I invite you, for example, to pick a study that caught your attention in the previous chapters and ask: "Why did the researchers choose this particular tool, and not another?" And, "How might another methodology contribute to this question?" Consider also how a particular experiment is crafted to fit with the constraints of a given tool (for example, by carefully controlling stimulus timing with MEG or EEG).

2. *Describe linking hypotheses that connect brain signals to linguistic representations*

Table 1 back on page 3 invited us to consider three different levels of description of a complex cognitive system like language. We might ask about the computational goals of that system, the algorithm by which those goals are reached, or the physical implementation of that algorithm. Neurolinguistics, we saw, demands our attention to all three levels and, importantly, how they connect together. These connections are linking hypotheses. Examples of linking hypotheses have been threaded through the book, although they perhaps have not always been clearly named as such.

One kind of linking hypotheses we have confronted concerns the signals that our tools measure. When a researcher reports some particular observation or result, they commit, whether explicitly or implicitly, to a particular linking hypothesis. On the relatively straightforward side are results of the sort "BOLD signal in the inferior frontal gyrus increased for condition A compared to condition B"; this entails a linking hypothesis along the lines of "The linguistic contents of condition A make additional hemodynamic demands at a certain moment in time compared to condition B." (While hypotheses and results are similar, note that the hypothesis is a statement about how things in the world connect together, while in contrast the results are a statement about fixed observations.) A slightly more complicated linking hypothesis might be something like "BOLD signal will increase logarithmically as the number of phrases in the sentence increase"[1] or perhaps "Speech signals with different frequencies are processed by neural populations that follow a spatial gradient."[2]

When linking hypotheses are laid out in this way, we see where they are incomplete and deserving of further refinement. Careful discussion of the linking hypotheses concerning the P600 ERP component, for example, pointed out a way to understand how it may be influenced by syntactic or semantic factors (as discussed around page 134 in Chapter 7) or highlighted gaps in understanding that must be addressed to disentangle how the brain may or may not decompose words into component morphemes (page 98 in Chapter 5). You can, and should, do this yourself with some of the many examples in the previous chapters. To take one of the examples from a moment ago: Why would BOLD signal increase logarithmically, and not linearly, for bigger phrase structures? Does this relationship hold for sentences that have long-distance dependencies between clauses? Or highly predictable sentences?

The bigger lesson here is that making linking hypotheses explicit and transparent is a necessary step in moving forward and making progress on difficult questions.

3. *Review state-of-the-art results that have emerged from this research*

Honestly, this is the easy part. There are a great many interesting and compelling lines of research, and fascinating and clever experiments designed to address them. It is easy for me, sometimes, to get drawn in to a "laundry list" of specific research studies, each with clever methods and complicated patterns of results. But it is necessary to combine close attention to the detail and nuance of individual studies with a broader perspective. That is, it is important to step back and see how individual efforts fit together coherently in order to contribute to addressing larger questions.

Taking such a step back involves asking several questions of the research, and of ourselves as researchers and students. Most obviously, we ask: "What is the goal of this study?" or some other variant that addresses the research question at issue. But I don't think this alone is enough to get a proper look at the bigger picture. Equally important is the question: "How does this study fit in with our current state of knowledge?" Perhaps you are reading a study that uses a new method to address a long-standing question? Or maybe previous research studies have shown conflicting results, and a new study sheds light on this conflict? There are a great many possible questions, and it would be a disservice to try to limit or constrain what counts as a valuable contribution to the broader scientific community. But my point is that we should pay attention to what that broader discussion might be, and in what way some particular research study fits.

In neurolinguistics, the state-of-the-art picture that emerges from this exercise still has, let's be honest, many holes. There is just a lot that we do not yet understand about how the brain implements the human faculty for language. This partial understanding manifests, for example, in the many conflicts and debates that we saw: to pick just two examples. In terms of the functional role of specific brain regions, like the inferior frontal gyrus, or the cognitive representation of linguistic building blocks such as morphemes. Each of these debates presents a truly difficult puzzle, but they offer some of the best opportunities to hone our abilities for critical thinking and careful scientific reasoning.

4. *Provide a foundation for you to read this literature for yourself*

It is my hope that the material reviewed here provides enough foundation that the interested reader could begin to engage with the primary research literature. This hope can only be put to the test when you, reader, follow the references and notes sprinkled through each chapter and start to read the studies that most capture your attention.

The thing is, I don't think you can expect to get the most out of a scientific research paper by picking it up and simply reading it from start to finish. In my own experience, I've found it helpful to read "non-linearly", that is, to skim some portions of a new paper before going back and reading more thoroughly. The idea here is to gather a good sense of what the larger story of the paper is: What are the researchers trying to do and what strategy are they using? Then, when I go in to make sense of the background research, the methods section, or results, my goal is to fit each particular detail into the bigger "story" that is being told in the paper. If a detail doesn't fit in (What does this term mean? Why did they use that method for this question?), then I know what parts I need to pay even more attention to.

Now, I have a very particular way I do this, and I wouldn't expect my strategy to work equally well for others. But, at least as a starting point, I tend to read in the following way: abstract (twice), conclusion section, abstract (again), all figures and tables, and then start a "proper" reading from the beginning. I encourage you to reflect on what sort of strategy might make it easiest to get a handle on the take away message of a paper, and then see how the nitty-gritty details serve to support that message.

5. *Point you towards the resources they may use to engage with this research themselves*

Follow the references and notes. Have I said that enough? Not only do they point to studies where you will learn more about this material, but they also point towards resources that you yourself can use as a neurolinguist. These resources include openly available repositories of data as well as free, often open-source research software that spans the many data types and methods used in neurolinguistics.[3,4]

Where are we (in terms of this field of research)?

Neurolinguistics is still a relatively young field. While it builds on pioneering work going back over 150 years, some of the most crucial tools for new insights, such as fMRI, are only a few decades old. The beginnings of the field were contemporaneous with the beginnings of a new kind of neuroscience: the localizationist perspective that specific functions of the mind were housed in specific structures or regions of the brain. Our modern understanding of the brain bases of language owes a great deal to this perspective, as evidenced by the maps throughout this book that localize specific linguistic computations and representations to locations in the brain as well as particular time-windows of processing; the *where* and *when* of processing.

Yet we also have seen that a fuller answer for *how* the brain creates language demands more. One important challenge is to engage with the notion that both language itself, and neural activity at the bases of how the brain works, are *dynamic*. Research has only begun to take on the massive challenge of developing linking hypotheses that are, likewise, fundamentally dynamic. A growing perspective in the neurosciences more broadly recognizes that such dynamics are embedded in interacting networks. On this view, the core cognitive building-blocks can only be found, not by identifying specific regions or time-points, but by teasing apart and unpacking the operations of such networks.[5] Indeed, we are starting to see the outlines of such interacting networks in speech perception, lexical processing, and sentence understanding, but we are certainly just at the beginning of this new stage.

One neural mechanism for implementing such dynamics may be neural oscillations, which we have seen crop up from time to time in the previous chapters. I am keen to see these lines of research develop further, as they offer one avenue to connect the macro-scale localizationist data that has been collected so far with more micro-scale facts about how neurons operate and interact with each other.

Another trend affecting the neurosciences more broadly, and neurolinguistics specifically, is the absolute explosion of data made possible by the many techniques that are now at our disposal. To give just one example, the Pubmed scientific database maintained by the National Institutes of Health[6] indexed just about 28,000 papers related to fMRI in 2010, but by 2018 that number was over 40,000. This increase in data must be met in kind with the development and refinement of theories and models into which these data fit; these are the linking hypotheses we dwelt on just above.

An area where more attention to data is needed concerns neurolinguistic studies from a broader array of languages. While there are over 7,000 languages spoken across the world, a very tiny fraction of those languages are represented in the studies discussed in this book. Differences between languages mean that some are more suitable than others for studying specific neurolinguistic research questions than others (see, for example, the discussions of Turkish vowels on page 64, Arabic morphology on page 96, and Japanese grammar on page 129). Moreover the languages that have been most studied tend to reflect existing sociopolitical power structures which leave many communities marginalized. Broadening the dataset that undergirds neurolinguistics to include a representative range of languages and language users is a necessary step towards developing more generalizable research.[7] Such steps build on existing cross-linguistic work that has focused on key topics such as multilingualism and sign languages which have received only glancing attention in these pages.

A theme that runs through all of these different trends is that of connection: connections between different brain networks, connections between data and models, or connections between data, models, and the people that make up our many different and varied language communities. It is perhaps too easy to also note that neurolinguistics is, of course, fundamentally about the connection between the scientific disciplines of neuroscience and linguistics. As an interdisciplinary science, researchers in neurolinguistics have seen directly how challenging it can be to make connections that actually work. Indeed, as an interdisciplinary science, neurolinguistics generates fierce debates as to whether and in what way insights from linguistics are "relevant" for studies of the brain, and in which ways studies of the brain do or do not "matter" for an understanding of the human capacity for language.[8] As fundamental a question as this is, it is almost certainly one that does not have a clear answer. There is no one way different scientific disciplines connect, one path to discover how the brain works, or one best tool for studying the human capacity for language. But that doesn't offer a free pass to go our separate ways. Rather, the challenge of interdisciplinary work is to make the connections between different ideas as explicit and as clear as possible – to specify the linking hypotheses that we have returned to again and again.

Call to action

As promised, this book is as much about the *questions* that we ask in neurolinguistics as it may be about any (partial) *answers* that scientists have identified thus far. These questions mean that the state of neurolinguistics is in flux, and this is exciting. It's exciting because the flux is grounded in a set of firm observations, including observations about *where* and *when* certain aspects of language are processed in the brain. It's from this foundation that exciting ideas and developments are emerging about *how* these systems operate together, interactively and dynamically, in the service of language.

Some of these developments will refine our current understanding, while others will most likely open the door to totally new perspectives and insights. The thing is: These opportunities are there, ready to be studied. We have so many good questions ready to be asked, and a range of methodologies and tools that can be mastered with a little bit of dedication, focus, and time. Don't just take my word for it. Take a look at the program at the next annual meeting of the Society for the Neurobiology of Language,[9] or visit your library to check out the latest issue of *Brain & Language*. I hope the book spurs your interest and leads to you to discover more about the research currently underway. In time, it may well lead you on a path to your own innovative discoveries.

Abbreviations

AG	Angular Gyrus
ASD	Autism Spectrum Disorders
ATL	Anterior Temporal Lobe
BOLD	Blood Oxygenation-Level Dependent
DCS	Direct Cortical Stimulation
DTI	Diffusion Tensor Imaging
ECoG	Electrocorticography
EEG	Electroencephalography
ERF	Event-Related Field
ERP	Event-Related Potential
fMRI	Functional Magnetic Resonance Imaging
fNIRS	Functional Near-Infrared Spectroscopy
IFG	Inferior Frontal Gyrus
IPL	Inferior Parietal Lobe
LATL	Left Anterior Temporal Lobe
LIFG	Left Inferior Frontal Gyrus
LPTL	Left Posterior Temporal Lobe
MMR	Mismatch Response
MRI	Magnetic Resonance Imaging
MVPA	Multi-Voxel Pattern Analysis
PET	Positron Emission Tomography
pMTG	Posterior Middle Temporal Gyrus
PPA	Primary Progressive Aphasia
PWD	Pure Word Deafness
rTMS	Repeated Transcranial Magnetic Stimulation
SMG	Supramarginal Gyrus
TMS	Transcranial Magnetic Stimulation
VBM	Voxel-based Morphometry
VLSM	Voxel-based Lesion Symptom Mapping
VMPFC	Ventro-Medial Prefrontal Cortex

Glossary of terms

agnosia A neural deficit that affects one's ability to recognize objects, sounds, or other perceptual categories.

anaphora Words that have their meaning filled in by context (for example, the pronoun "she").

anomia A kind of aphasia that involves difficulty accessing words from the mental lexicon.

aphasia A language deficit caused by brain damage.

classical model Early account of the neural bases of language that focused on language comprehension in the left posterior temporal lobe, and language production in the left inferior frontal lobe.

cytoarchitecture The arrangement of neurons into layers in the cortex of the brain; used to differentiate brain regions.

diffusion tensor imaging (DTI) Method to measure the structural connectivity between brain regions by measuring the diffusion of water along axons using MRI.

electrocorticography (ECoG) Method that invasively measures neural voltage potentials in patients undergoing neurosurgery using electrodes placed directly onto the surface, or penetrating into, cortical tissue with high spatial and temporal resolution.

electroencephalography (EEG) Method to measure scalp voltage potentials generated by tens of thousands of cortical neurons with high temporal but low spatial resolution.

event-related potential (ERP) Time-aligned average of EEG signals used to amplify cortical signals associated with perceptual or cognitive processes.

functional magnetic resonance imaging (fMRI) Method to measure neural activity through the proxy of blood oxygenation with high spatial resolution but low spatial resolution.

linking hypothesis (or **linking function**) Account specifying how computations or representations at one level of description are connected to those at another (for example, a specification of how recognizing a phoneme might impact brain activity recorded in a particular experimental setting).

M100 Evoked neuromagnetic signal recorded with MEG that reflects activation in the auditory cortex about 100 milliseconds after sound begins.

magnetic resonance imaging (MRI) Medical imaging tool used to create high-resolution three-dimensional images of tissues, including the brain, by recording the energy released by hydrogen atoms that are perturbed by a strong magnetic field.

magnetoencephalography (MEG) Method to measure very small magnetic fields generated by cortical neural activity with high temporal resolution and moderate spatial resolution.

mismatch response (MMR) Evoked neural signal, measurable with EEG or MEG, which is observed when a sequence of similar stimuli (the "standards") are interrupted by something with a different property (the "deviant"). Used to study the processing and structure of categories, including phonemes.

N400 Event-related potential with a negative voltage peak approximatly 400 milliseconds after stimulus onset on the central posterior area of the scalp. It is evoked by stimuli with a meaning that is unexpected or unlikely.

neural oscillations Rhythmic neural activity at discrete frequencies thought to be important for synchronizing diverse populations of neurons to facilitate information processing.

neurogram Neural representation of continuous acoustic information including frequency spectra and temporal dynamics.

P600 Event-related potential with a positive voltage peak approximately 600 milliseconds after stimulus onset on posterior areas of the scalp. It is most often evoked by stimuli that are ungrammatical or syntactically unexpected.

periodotopy Spatial organization of neurons found in the auditory cortex such that adjacent populations of neurons respond to sounds with adjacent temporal dynamics.

primary progressive aphasia (PPA) Family of neurodegenerative disorders that affect language comprehension and production.

semantic hub Area of the anterior temporal lobes of the brain that has been associated conceptual knowledge.

temporal windows of integration Account of the neural bases of speech perception that is based the perceptual sampling of acoustic information in discrete windows, perhaps via the mechanism of neural oscillations.

tonotopy Spatial organization of neurons such that adjacent populations of neurons respond to adjacent sound frequencies.

transcranial magnetic stimulation (TMS, rTMS) Method to temporarily excite or inhibit neural activity from a small area using a magnetic field generated outside of the scalp.

word embedding Computational technique to represent the meaning of the word as vector describing a point in a semantic space whose dimensions may be other words or fixed concepts.

International Phonetic Alphabet for English

The International Phonetic Alphabet, or IPA, is maintained by the International Phonetic Association as a resource to write down the speech sounds of the world's languages. Below is a quick guide to a subset of the IPA that is used in this book, following the author's dialect of American English. (This table has been adapted from Fromkin et al., 2017.)

p	pill	j	you
t	till	ʃ	shill
k	kill	w	witch
b	bill	ʒ	measure
d	dill	i	beet
g	gill	ɪ	bit
m	mill	e	bait
n	nil	ɛ	bet
ŋ	ring	u	boot
f	feel	ʊ	foot
s	seal	o	boat
h	heal	ɔ	bore
v	veal	æ	bat
z	zeal	a	pot
l	leaf	ʌ	butt
θ	thigh	ə	sofa
tʃ	chill	aɪ	bite
ɹ	reef	aʊ	bout
ð	thy	ɔɪ	boy
dʒ	gin		

Notes

Chapter 1. Introduction

1. The introduction in Marr's 1982 book *Vision* offers an accessible guide to this way of thinking about cognitive systems.
2. In fact such work earned Karl von Frisch the 1973 Nobel Prize in Physiology or Medicine (von Frisch, 1974).
3. See Poeppel and Embick (2005); Embick and Poeppel (2015) for a more in-depth discussion of these challenges with a focus on language.
4. For reviews of this line of research, see Knudsen (2002); Peña and DeBello (2010)
5. If the syntactic diagrams in Fig. 2 look unfamiliar, this may be a good moment to get your hands on an Introductory Linguistics textbook or other similar resource. Note 6 has some suggestions.
6. This book does not assume that you have any familiarity or background with neuroscientific methods. However, it does assume a basic familiarity with linguistics. There are many excellent introductions to linguistics. Two books that I use in my own teaching are Adger (2019) and Fromkin et al. (2017). There are also many accessible multi-media resources, such as the wonderful YouTube videos on the *Lingthusiasm* channel created by Gretchen McCulloch and Lauren Gawne.
7. Another great re-telling is found in Carl Sagan's 1979 collection of essays titled, appropriately enough, *Broca's Brain*.
8. Signoret et al. (1984) offer more details about the (re)discovery of Leborgne's brain.
9. Parker Jones et al. (2018) offer a modern perspective on the methods of phrenology.
10. The views of phrenology were more than just nonsense, in fact, as there is a long and deeply troubling history in medicine and science of linking physiological characteristics, such as the shape of the cranium, with racist practices and policies (e.g. Gould, 1981).
11. Uttal (2003) offers a critical perspective on localizationist approaches in modern neuroscience.

12. Miller (2003) presents an enjoyable first-person perspective on what has come to be called "The Cognitive Revolution." Chomsky (1965) is a key contribution from the perspective of language.
13. The levels of description were quite directly influenced by Chomsky's distinction between *linguistic competence* and *performance* (Marr, 1982, p. 28).

Chapter 2. The toolbox

1. The left side of an *axial* image is not always the left hemisphere of the brain. Under the "radiological" convention, the left side of the brain is shown on the right, while under the "neurological" convention, left is shown on left. My advice: always look for annotations to guide you to the orientation of any brain image.
2. There are a number of interactive brain atlases on the internet, for example: https://www.humanbrainproject.eu/en/explore-the-brain/ (accessed June 1 2021). Check them out and explore!
3. There are many great introductory resources for cellular neuroscience out there, including the interactive *Get Body Smart* ebook: https://www.getbodysmart.com/nerve-cells (accessed June 1 2021).
4. Case studies from Warrington and McCarthy (1983) and Warrington and Shallice (1984).
5. Logothetis and Wandell (2004) offer a nuanced discussion of how this hemodynamic response relates to neuronal activity.
6. Another mark against the temporal resolution of fMRI is the time it takes to record each snapshot of BOLD signal (the sampling rate). Typical recording protocols record a full 3D image every 2 seconds (TR, or repetition time, equals 2 seconds). While this is pretty common, there exist protocols that can take images at faster rates, even down to less than half a second (Feinberg et al., 2010). So, the principal reason for fMRI being a "slow" technique is really the sluggish hemodynamic response.
7. Poldrack et al. (2011) provide a comprehensive introduction to fMRI data and analysis.
8. Indeed, the first EEG measurements were made in the 1920s by the psychiatrist Hans Berger.
9. Luck (2014) offers a very accessible introduction to EEG and ERPs.
10. Cohen (2014) provides a comprehensive introduction to different techniques for analyzing electrophysiological data.
11. See Baillet et al. (2001) and Hämäläinen et al. (1993) for in-depth introductions to MEG.
12. Hallett (2007) provides an overview of TMS.
13. Ojemann et al. (1989) present a detailed look at what the invasive DCS technique can reveal about language.
14. See Hartwigsen and Saur (2019) for a review focused on aphasia recovery.

Chapter 3. Sounds in the brain

1. To get started with sign language neurolinguistics, take a look at Emmorey and Ozyurek (2014); MacSweeney et al. (2008); Corina and Blau (2016).
2. Ringach (2004) reviews research on receptive fields in the visual cortex with an historical perspective.
3. Shannon et al. (1995). You can sample some noise vocoded speech yourself at http://www.mrc-cbu.cam.ac.uk/personal/matt.davis/vocode/ (accessed Dec 21 2021).
4. See Lago et al. (2015) for categorical perception of fricatives, and references cited therein for literature on categorical perception in speech more broadly.
5. See, for example, the work of Munson (2011).
6. This idea was first introduced in Poeppel (2003); see Giraud and Poeppel (2012a) for a comprehensive discussion.
7. Giraud et al. (2007); see also Morillon et al. (2012).

Chapter 4. A neural code for speech

1. This "balance" is not the case for English, for example, where all rounded vowels like [o] and [u] are also back vowels.
2. See Roberts et al. (2000) for a review of the M100 evoked response.
3. See also the review by Yi et al. (2019).
4. Another wonderful window into "phonotopy" comes from the rare and remarkable single-neural data in humans reported by Chan et al. (2014).
5. Haxby et al. (2014) offer an introduction and review of MVPA.
6. See Clements and Hume (1995) for an in-depth take on the organization of phonological features.
7. In EEG, the mismatch response is a negative-going potentials called the "mismatch negativity" or *MMN*. In MEG, the same response is called the "mismatch field" or *MMF*.
8. For evidence linking the MMR with phonological differences, see Näätänen et al. (1997); Phillips et al. (2000).
9. Also see Eulitz and Lahiri (2004)
10. See Polster and Rose (1998) and Poeppel (2001) for more discussion.
11. Rizzolatti and Arbib (1998) develop one account that takes the motor system to be a central component in perception; Gregory Hickok's 2014 book, *The Myth of Mirror Neurons*, offers an in-depth and critical perspective.
12. This will be one of the few times I discuss "language dominance" in the book. Though the idea that one hemisphere is "responsible" for language captures the public imagination, it's simply not correct. We'll see throughout the book

that regions from both the left and right hemispheres are involved in different aspects of language.
13. Groen et al. (2008) and Kelley (2011) offer overviews of how language is affected in ASD.
14. See Kenet (2011) for an overview of sensory perception aspects of ASD.
15. See Kujala et al. (2013) for extensive discussion of the possible link between the auditory processing of speech and language-related deficits in ASD.

Chapter 5. Activating words

1. You can try this out yourself here: https://www.psytoolkit.org/experiment-library/ldt.html; the phenomenon was first reported by Meyer and Schvaneveldt (1971) (accessed Dec 21 2021).
2. Kravitz et al. (2013) review the ventral visual pathway.
3. These sign language data come from Leonard et al. (2012).
4. For a broader discussion of the brain as a "prediction engine," see Bar (2011).
5. See Gagnepain et al. (2012) for computational simulations showing how these kind of predictions facilitate processing.
6. Also see evidence from spoken word recognition from Gagnepain et al. (2012).
7. Lewis and Poeppel are building on work on visual word recognition by, for example, Solomyak and Marantz (2009) and Hauk et al. (2006).
8. A note on terminology: The plural "-s" marker is an *inflectional* suffix as it doesn't change the word's category ("cat" and "cats" are both nouns). In contrast, "-al" is an example of a *derivational* suffix in that it changes the word from one category to another.
9. This discussion is drawn from the review by Pinker and Ullman (2002).
10. See also Gagnepain et al. (2012). For related work on irregular morphology in English, see Fruchter et al. (2013).

Chapter 6. Representing meaning

1. The quotation comes from Wernicke (1977, p. 179), as cited by Gage and Hickok (2005, p. 824).
2. The examples here of category-specific agnosia come from Warrington and Shallice (1984).
3. Mitchell et al. (2008) first demonstrated how word embeddings can be used to decode neural activity.
4. Patterson et al. (2007) offer a comprehensive review of semantic dementia.

5. Patterson and Lambon Ralph (2016) review evidence for the *distributed-plus-hub* theory in more detail. Rogers et al. (2004) offer a computational implementation of this theory.
6. These case studies come from Rumiati et al. (2001); see Mahon and Caramazza (2005) for further discussion.
7. For a deeper dive into this debate, see also Mahon and Caramazza (2008) and the dialogue between Gregory Hickok and Corrado Sinigaglia in Hickok and Sinigaglia (2013).

Chapter 7. Structure and prediction

1. For reviews of research on the N400, see Kutas and Federmeier (2000); Lau et al. (2008); Kutas and Federmeier (2011).
2. See e.g. Bentin et al. (1985) for semantic priming effects on the N400; word frequency effects were first discussed by Van Petten and Kutas (1990).
3. Laszlo and Federmeier (2009) present a similar study using pseudowords.
4. See also Lau et al. (2016)
5. Ferreira and Patson (2007) offer an overview of this "good enough" approach to language comprehension.
6. Indeed, Ehrenhofer et al. (To appear) describe a set of EEG results that appear to conflict with, or at least demand modifications of, the theory sketched on page 129.
7. That same study also found another ERP component for the comparison illustrated in example (7) that they called the *early left anterior negativity* or ELAN. Studies have linked the ELAN with early, perhaps predictive, analysis of grammatical information. See Steinhauer and Drury (2012) and Lau et al. (2006) for discussion.
8. For reviews and discussion, see Van Petten and Luka (2012); Gouvea et al. (2010); Brouwer et al. (2017).
9. Kuperberg (2007) offers an alternative interpretation of the "semantic" P600 to the prediction-based account presented here.

Chapter 8. Composing sentences

1. See Mazoyer et al. (1993) and Stowe et al. (1998). The examples in (1) come from from Humphries et al. (2006).
2. Indeed, after hearing the poem, Alice herself remarks:
   > Somehow it seems to fill my head with ideas – only I don't exactly know what they are! However, somebody killed something: that's clear, at any rate

   (*Through the Looking-Glass*, Chapter 2).

3. See Pylkkänen et al. (2008) for possible interpretations of the role of the VMPFC during sentence comprehension.
4. These MEG findings stand in interesting contrast with findings from alternative techniques like fMRI. Studies using the latter technique, even using a similar "simple composition" protocol, have pointed towards involvement of the LIFG (Zaccarella and Friederici, 2015; Zaccarella et al., 2017a). One factor to consider when confronted with apparently conflicting evidence such as this is temporal resolution: Whereas MEG reflects immediate and transient neural responses to language, the fMRI signal is most sensitive to neural activation that is sustained over several seconds.
5. Blanco-Elorrieta et al. (2018); see Pylkkänen (2016) for a review.
6. Wilson et al. (2014) discusses the syntactic deficits (or lack thereof) of patients with damage in the LATL.
7. See Hillis et al. (2017) for an interesting discussion of the pitfalls and possible solutions when deficit/lesion and correlational data don't appear to line up.
8. Brennan (2016) reviews the pros and cons of naturalistic experiments for neurolinguistics.

Chapter 9. Building dependencies

1. Caramazza and Zurif (1976) explore this effect across several groups of aphasic patients; see also Caramazza and Berndt (1978) for a review.
2. Grodzinsky et al. (1999); Berndt and Caramazza (1999); Zurif and Pinango (1999) provide a back-and-forth discussion of the challenges facing efforts to generalize across aphasia patients.
3. See Gorno-Tempini et al. (2004) for an overview of PPA.
4. See also Wilson et al. (2012) for a broader review of PPA and syntax.
5. Rogalsky and Hickok (2010) offer a critical overview of several leading hypotheses.
6. Fiebach et al. (2005) present one version of a working-memory account of the LIFG.
7. Psycholinguistic evidence for this sort of prediction comes from the "filled-gap effect" discussed by, for example, Stowe (1986) and Phillips (2006).
8. There is quite a bit of evidence that anaphora are based on different syntactic representations than *wh*-question, passive sentences, and other such constructions. For example, you can string pronouns and nouns together across sentences and even a large discourse. Still, some syntactic theories do suggest that similar syntactic rules might be involved (e.g. Hornstein, 1999).
9. For fMRI research on agreement dependencies, see e.g. Carreiras et al. (2015); for deficit/lesion research, see e.g. Barbieri et al. (2021).

10. See, for example, Nelson et al. (2017) in addition to studies like those by Wehbe et al. (2014); Brennan et al. (2016); Lopopolo et al. (2017) already discussed earlier.
11. Meyer et al. (2019) offer an overview of this work. See also Bastiaansen and Hagoort (2006); Ding et al. (2016); Meyer (2018).

Chapter 10. Wrapping up

1. As suggested by Pallier et al. (2011)
2. As reported by Barton et al. (2012)
3. There are growing efforts to openly share research data. Two resources that are especially valuable for neurolinguistics are the OpenNeuro project at https://openneuro.org and the Open Science Foundation at https://osf.io.
4. Many excellent tools can be found via the NeuroImaging Tools & Resources Collaboratory at https://www.nitrc.org.
5. The fundamental role of networks in neuroscience is discussed by Sporns (2016).
6. https://pubmed.ncbi.nlm.nih.gov/.
7. The importance of linguistic diversity for neurolinguistics is highlighted by Bornkessel-Schlesewsky and Schlesewsky (2016).
8. See, for example, the debate "What counts as neurobiology of language?" hosted at the 2014 meeting of the Society for the Neurobiology of Language.
9. https://www.neurolang.org.

Figure acknowledgments

Figure 3. Over 150 years of language in the brain A: Reprinted from Dronkers, N. F., Plaisant, O., Iba-Zizen, M. T., and Cabanis, E. A. (2007). Paul Broca's historic cases: High resolution MR imaging of the brains of Leborgne and Lelong. *Brain*, 130(Pt 5):1432–41, by permission of Oxford University Press.
B: Reprinted from Wernicke, C. (1874). *Der aphasische Symptomencomplex: eine psychologische Studie auf anatomischer Basis*. Cohn & Weigert.
C: From Geschwind, N. (1970). The organization of language and the brain. *Science*, 170(3961):940–4. Reprinted with permission from AAAS.
D: Adapted from Friederici, A. D. (2012). The cortical language circuit: From auditory perception to sentence comprehension. *Trends Cogn. Sci.*, 16(5):262–8, ©Elsevier 1988, with permission from Elsevier.

Figure 7. The deficit/lesion method A: Reprinted from Dronkers, N. F., Plaisant, O., Iba-Zizen, M. T., and Cabanis, E. A. (2007). Paul Broca's historic cases: High resolution MR imaging of the brains of Leborgne and Lelong. *Brain*, 130(Pt 5):1432–41, by permission of Oxford University Press.
B: Reprinted from Dronkers, N. F., Redfern, B. B., and Knight, R. T. (1999). The neural architecture of language disorders. In Gazzaniga, Michael, ed., *The New Cognitive Neurosciences, second edition*, p. 953, ©1999 Massachusetts Institute of Technology, by permission of The MIT Press.
C: Adapted from Rogalsky, C., Love, T., Driscoll, D., Anderson, S. W., and Hickok, G. (2011). Are mirror neurons the basis of speech perception? Evidence from five cases with damage to the purported human mirror system. *Neurocase*, 17(2):178–87, by permission of Taylor & Francis. www.tandfonline.com.

Figure 12. The neurogram A: From Kubanek, J., Brunner, P., Gunduz, A., Poeppel, D., and Schalk, G. (2013). The tracking of speech envelope in the human cortex. *PLoS One*, 8(1):e53398. Kubanek et al. (2013). Used under a CC BY 4.0 International license. https://creativecommons.org/licenses/by/4.0/
B. and C: From Barton, B., Venezia, J. H., Saberi, K., Hickok, G., and Brewer, A. A. (2012). Orthogonal acoustic dimensions define auditory field maps in

human cortex. *Proceedings of the National Academy of Sciences of the United States of America*, 109(50):20738–43.

Figure 14. A dual-stream model for speech perception Adapted with permission from Springer Nature from Hickok, G. and Poeppel, D. (2007). The cortical organization of speech processing. *Nat Rev Neurosci.*, 8(5):393–402. ©Springer Nature 2007.

Figure 15. The neural representation of phonemes A and B: Reprinted from Scharinger, M., Idsardi, W. J., and Poe, S. (2011). A comprehensive three-dimensional cortical map of vowel space. In *J Cogn Neurosci*, 23(12), pp. 3972–82. ©2011 Massachusetts Institute of Technology, by permission of The MIT Press.
C: From Mesgarani, N., Cheung, C., Johnson, K., and Chang, E. F. (2014). Phonetic feature encoding in human superior temporal gyrus. *Science*, 343(6174):1006–10. Reprinted with permission from AAAS.
D: From Arsenault, J. S. and Buchsbaum, B. R. (2015). Distributed neural representations of phonological features during speech perception. *Journal of Neuroscience*, 35(2):634–42. Used under a CC BY 4.0 International license. https://creativecommons.org/licenses/by/4.0/

Figure 16. Voxel-based lesion symptom mapping (VLSM) Top: Adapted from Baldo, J. V., Wilson, S. M., and Dronkers, N. F. (2012). Uncovering the Neural Substrates of Language: A Voxel-Based Lesion-Symptom Mapping Approach. In Faust, M., editor, *The Handbook of the Neuropsychology of Language*, pages 582–594. With permission from Wiley-Blackwell Bottom: Adapted by permission from Springer Nature from Bates, E., Wilson, S. M., Saygin, A. P., Dick, F., Sereno, M. I., Knight, R. T., and Dronkers, N. F. (2003). Voxel-based lesion-symptom mapping. *Nat Neurosci*, 6(5):448–50. ©Springer Nature 2003.

Figure 17. The time course and localization of lexical activation A: Adapted from Marinkovic, K., Dhond, R. P., Dale, A. M., Glessner, M., Carr, V., and Halgren, E. (2003). Spatiotemporal dynamics of modality specific and supramodal word processing. *Neuron*, 38(3):487–97, ©Elsevier 2003, with permission from Elsevier.
B: Adapted from Leonard, M. K., Ferjan Ramirez, N., Torres, C., Travis, K. E., Hatrak, M., Mayberry, R. I., and Halgren, E. (2012). Signed words in the congenitally deaf evoke typical late lexicosemantic responses with no early visual responses in left superior temporal cortex. *Journal of Neuroscience*, 32(28):9700–5. Used under a CC-BY-NC-SA 3.0 license. https://creativecommons.org/licenses/by-nc-sa/3.0/

FIGURE ACKNOWLEDGMENTS

Figure 18. Stages of spoken-word recognition Adapted from Lewis, G. and Poeppel, D. (2014). The role of visual representations during the lexical access of spoken words. *Brain Lang.*, 134(0):1–10, ©Elsevier 2014, with permission from Elsevier.

Figure 19. The semantic system A: Reprinted from Binder, J. R., Desai, R. H., Graves, W. W., and Conant, L. L. (2009). Where is the semantic system? A critical review and meta-analysis of 120 functional neuroimaging studies. *Cerebral Cortex*, 19(12):2767–9, by permission of Oxford University Press.
B: Adapted by permission from Springer Nature from Huth, A. G., de Heer, W. A., Griffiths, T. L., Theunissen, F. E., and Gallant, J. L. (2016). Natural speech reveals the semantic maps that tile human cerebral cortex. *Nature*, 532(7600):453–8. ©Springer Nature 2016.

Figure 20. Semantic dementia Reprinted by permission from Springer Nature from Patterson, K., Nestor, P. J., and Rogers, T. T. (2007). Where do you know what you know? The representation of semantic knowledge in the human brain. *Nat Rev Neurosci*, 8(12): 976–987. ©Springer Nature 2007.

Figure 21. Components of sentence structure The stock photo in Fig 21 should be attributed to Marco Verch, used under the CC BY 2.0 license. https://creativecommons.org/licenses/by/2.0/

Figure 22. The N400 event-related-potential A: From Kutas, M. and Hillyard, S. A. (1980). Reading senseless sentences: Brain potentials reflect semantic incongruity. *Science*, 207:203–205. Reprinted with permission from AAAS
B: Reprinted from Kutas, M. and Federmeier, K. D. (2011). Thirty years and counting: Finding meaning in the N400 component of the event-related brain potential (ERP). *Annual Review of Psychology*, Vol 62, 62:621–47.
C: Reprinted from Holcomb, P. J. (1988). Automatic and attentional processing: An event-related brain potential analysis of semantic priming. *Brain and Language*, 35(1):66–85, ©Elsevier 1988, with permission from Elsevier.
D: Reprinted from Holcomb, P. J. and McPherson, W. B. (1994). Event-related brain potentials reflect semantic priming in an object decision task. *Brain and Cognition*, 24(2):259–76, ©Elsevier 1994, with permission from Elsevier.
E: Reprinted from Federmeier, K. D. and Kutas, M. (1999). A rose by any other name: Long-term memory structure and sentence processing. *J. Mem. Lang.*, 41(4):469–95, ©Elsevier 1999, with permission from Elsevier.

FIGURE ACKNOWLEDGMENTS 189

Figure 23. Predictability and the N400 A: From Nieuwland, M. S., Barr, D. J., Bartolozzi, F., Busch-Moreno, S., Darley, E., Donaldson, D. I., Ferguson, H. J., Fu, X., Heyselaar, E., Huettig, F., Husband, E. M., Ito, A., Kazanina, N., Kogan, V., Kohút, Z., Kulakova, E., Mézière, D., Politzer-Ahles, S., Rousselet, G., Rueschemeyer, S.-A., Segaert, K., Tuomainen, J., and Von Grebmer Zu Wolfsthurn, S. (2019). Dissociable effects of prediction and integration during language comprehension: Evidence from a large-scale study using brain potentials. *Philos. Trans. R. Soc. Lond. B. Biol. Sci.* Reused with permission from the authors.

B: Reprinted from van Berkum, J. J. A., Zwitserlood, P., Hagoort, P., and Brown, C. M. (2003). When and how do listeners relate a sentence to the wider discourse? Evidence from the N400 effect. *Cogn. Brain Res.*, 17(3):701–18, ©Elsevier 2003, with permission from Elsevier.

C: Reprinted from Van Berkum, J. J. A., van den Brink, D., Tesink, C. M. J. Y., Kos, M., and Hagoort, P. (2008). The neural integration of speaker and message. In *J Cogn Neurosci*, 20(4), pp. 580–91. ©2008 Massachusetts Institute of Technology, by permission of The MIT Press.

D: From Brennan, J. R. and Hale, J. T. (2019). Hierarchical structure guides rapid linguistic predictions during naturalistic listening. *PLoS ONE*, 14(1):e0207741. Used under a CC BY 4.0 International license. https://creativecommons.org/licenses/by/4.0/

Figure 24. The P600 ERP Reprinted from Gouvea, A. C., Phillips, C., Kazanina, N., and Poeppel, D. (2010). The linguistic processes underlying the P600. *Lang. Cogn. Process.*, 25(2):149–88. Taylor & Francis. www.tandfonline.com.

Figure 27. Prediction dynamics Adapted from Klimovich-Gray, A., Tyler, L. K., Randall, B., Kocagoncu, E., Devereux, B., and Marslen-Wilson, W. D. (2019). Balancing prediction and sensory input in speech comprehension: The spatiotemporal dynamics of word recognition in context. *Journal of Neuroscience*, 39(3):519–27. Used under a CC BY 4.0 International license. https://creativecommons.org/licenses/by/4.0/

Figure 28. A combinatoric brain network A and B: From Pallier, C., Devauchelle, A.-D., and Dehaene, S. (2011). Cortical representation of the constituent structure of sentences. *Proc. Natl. Acad. Sci.*, 108(6):2522–27.

C and D: Reprinted from Zaccarella, E., Schell, M., and Friederici, A. D. (2017b). Reviewing the functional basis of the syntactic Merge mechanism for language: A coordinate-based activation likelihood estimation meta-analysis. *Neuroscience and Biobehavioral Reviews*, 80:646–56, ©Elsevier 2017, with permission from Elsevier.

Figure 29. Simple composition and the anterior temporal lobe A: From Bemis, D. K. and Pylkkänen, L. (2011). Simple composition: A magnetoencephalography investigation into the comprehension of minimal linguistic phrases. *J. Neurosci.*, 31(8):2801–14. Used under a CC BY-NC-SA 3.0 license. https://creativecommons.org/licenses/by-nc-sa/3.0/
B: From Blanco-Elorrieta, E., Kastner, I., Emmorey, K., and Pylkkänen, L. (2018). Shared neural correlates for building phrases in signed and spoken language. *Sci Rep*, 8(1):5492. Used under a CC BY 4.0 International license.
C: Reprinted from Zhang, L. and Pylkkänen, L. (2015). The interplay of composition and concept specificity in the left anterior temporal lobe: An MEG study. *Neuroimage*, 111:228–40, ©Elsevier 2015, with permission from Elsevier.

Figure 30. Syntax in the posterior temporal lobe A: Reprinted from Matchin, W., Brodbeck, C., Hammerly, C., and Lau, E. (2018). The temporal dynamics of structure and content in sentence comprehension: Evidence from fMRI-constrained MEG. *Human Brain Mapping*. With permission from Wiley-Blackwell.
B: From Frankland, S. M. and Greene, J. D. (2015). An architecture for encoding sentence meaning in left mid-superior temporal cortex. *Proceedings of the National Academy of Sciences of the United States of America*, 112(37):11732–7.

Figure 31. Dependency processing in the LIFG A: From Amici, S., Brambati, S. M., Wilkins, D. P., Ogar, J., Dronkers, N. L., Miller, B. L., and Gorno-Tempini, M. L. (2007). Anatomical correlates of sentence comprehension and verbal working memory in neurodegenerative disease. *Journal of Neuroscience*, 27(23):6282–90. Used under a CC BY 4.0 International license. https://creativecommons.org/licenses/by/4.0/
B: Reprinted from Matchin, W., Sprouse, J., and Hickok, G. (2014). A structural distance effect for backward anaphora in Broca's area: An fMRI study. *Brain and Language*, 138:1–11, ©Elsevier 2014, with permission from Elsevier.

References

Adger, D. (2019). *Language Unlimited: The Science Behind Our Most Creative Power.* Oxford: Oxford University Press.

Amici, S., Brambati, S. M., Wilkins, D. P., Ogar, J., Dronkers, N. L., Miller, B. L., and Gorno-Tempini, M. L. (2007). Anatomical correlates of sentence comprehension and verbal working memory in neurodegenerative disease. *Journal of Neuroscience*, 27(23):6282–90.

Arsenault, J. S. and Buchsbaum, B. R. (2015). Distributed neural representations of phonological features during speech perception. *Journal of Neuroscience*, 35(2):634–42.

Baillet, S., Mosher, J. C., and Leahy, R. M. (2001). Electromagnetic brain mapping. *IEEE Signal Processing*, 18(6):14–30.

Baldo, J. V., Wilson, S. M., and Dronkers, N. F. (2012). Uncovering the neural substrates of language: a voxel-based lesion–symptom mapping approach. In Faust, M., editor, *The Handbook of the Neuropsychology of Language*, pages 582–94. Oxford: Wiley-Blackwell.

Bar, M. (2011). *Predictions in the Brain: Using Our Past to Generate a Future.* Oxford: Oxford University Press.

Barbieri, E., Litcofsky, K. A., Walenski, M., Chiappetta, B., Mesulam, M.-M., and Thompson, C. K. (2021). Online sentence processing impairments in agrammatic and logopenic primary progressive aphasia: evidence from ERP. *Neuropsychologia*, 151:107728.

Barton, B., Venezia, J. H., Saberi, K., Hickok, G., and Brewer, A. A. (2012). Orthogonal acoustic dimensions define auditory field maps in human cortex. *Proceedings of the National Academy of Sciences of the United States of America*, 109(50):20738–43.

Bastiaansen, M. and Hagoort, P. (2006). Oscillatory neuronal dynamics during language comprehension. *Progress in Brain Research*, 159:179–96.

Bates, E., Wilson, S. M., Saygin, A. P., Dick, F., Sereno, M. I., Knight, R. T., and Dronkers, N. F. (2003). Voxel-based lesion–symptom mapping. *Nature Neuroscience*, 6(5):448–50.

Bemis, D. K. and Pylkkänen, L. (2011). Simple composition: a magnetoencephalography investigation into the comprehension of minimal linguistic phrases. *Journal of Neuroscience*, 31(8):2801–14.

Bentin, S., McCarthy, G., and Wood, C. C. (1985). Event-related potentials, lexical decision and semantic priming. *Electroencephalography and Clinical Neurophysiology*, 60(4):343–55.

Berndt, R. S. and Caramazza, A. (1999). How "regular" is sentence comprehension in Broca's aphasia? It depends on how you select the patients. *Brain and Language*, 67(3):242–47.

Binder, J. R., Desai, R. H., Graves, W. W., and Conant, L. L. (2009). Where is the semantic system? A critical review and meta-analysis of 120 functional neuroimaging studies. *Cerebral Cortex*, 19(12):2767–96.

Blanco-Elorrieta, E., Kastner, I., Emmorey, K., and Pylkkänen, L. (2018). Shared neural correlates for building phrases in signed and spoken language. *Scientific Reports*, 8(1):5492.

Blanco-Elorrieta, E. and Pylkkänen, L. (2016). Composition of complex numbers: delineating the computational role of the left anterior temporal lobe. *Neuroimage*, 124(Pt A):194–203.

Bornkessel-Schlesewsky, I. and Schlesewsky, M. (2016). The importance of linguistic typology for the neurobiology of language. *Linguistic Typology*, 20(3): 615–21.

Boynton, G. M., Engel, S. A., and Heeger, D. J. (2012). Linear systems analysis of the fMRI signal. *NeuroImage*, 62(2):975–84.

Brennan, J. R. (2016). Naturalistic Sentence Comprehension in the Brain. *Language and Linguistics Compass*, 10(7), 299–313. https://doi.org/10/gg6xcc

Brennan, J. R. and Hale, J. T. (2019). Hierarchical structure guides rapid linguistic predictions during naturalistic listening. *PLoS ONE*, 14(1):e0207741.

Brennan, J. R., Stabler, E. P., Van Wagenen, S. E., Luh, W.-M., and Hale, J. T. (2016). Abstract linguistic structure correlates with temporal activity during naturalistic comprehension. *Brain and Language*, 157-8:81–94.

Brodmann, K. (1909). *Vergleichende Lokalisationslehre der Grosshirnrinde in ihren Prinzipien dargestellt auf Grund des Zellenbaues*. Leipzig: Barth.

Brouwer, H., Crocker, M. W., Venhuizen, N. J., and Hoeks, J. C. J. (2017). A neurocomputational model of the N400 and the P600 in language processing. *Cognitive Science*, 41 Suppl. 6:1318–52.

Caramazza, A. and Berndt, R. S. (1978). Semantic and syntactic processes in aphasia: a review of the literature. *Psychological Bulletin*, 85(4):898–918.

Caramazza, A. and Zurif, E. B. (1976). Dissociation of algorithmic and heuristic processes in language comprehension: evidence from aphasia. *Brain and Language*, 3(4):572–82.

Carreiras, M., Quiñones, I., Mancini, S., Hernández-Cabrera, J. A., and Barber, H. (2015). Verbal and nominal agreement: an fMRI study. *NeuroImage*, 120:88–103.

Chan, A. M., Dykstra, A. R., Jayaram, V., Leonard, M. K., Travis, K. E., Gygi, B., Baker, J. M., Eskandar, E., Hochberg, L. R., Halgren, E., and Cash, S. S. (2014). Speech-specific tuning of neurons in human superior temporal gyrus. *Cerebral Cortex*, 24(10):2679–93.

Chomsky, N. (1965). *Aspects of the Theory of Syntax*. Cambridge, MA: MIT Press.

Chow, W.-Y., Smith, C., Lau, E., and Phillips, C. (2016). A "bag-of-arguments"mechanism for initial verb predictions. *Language and Cognitive Neuroscience*, 31(5):577–96.

Clements, G. N. and Hume, E. V. (1995). The internal organization of speech sounds. In Goldsmith, J., editor, *The Handbook of Phonological Theory*, pages 245–302. Oxford: Blackwell.

Cohen, M. X. (2014). *Analyzing Neural Time Series Data*. Cambridge, MA: MIT Press.
Corina, D. P. and Blau, S. (2016). Neurobiology of sign language. In Hickok, G. and Small, S. L., editors, *Neurobiology of Language*. New York: Academic Press.
Correia, J., Formisano, E., Valente, G., Hausfeld, L., Jansma, B., and Bonte, M. (2013). Brain-based translation: fMRI decoding of spoken words in bilinguals reveals language-independent semantic representations in anterior temporal lobe. *Journal of Neuroscience*, 34(1):332.
Crinion, J., Turner, R., Grogan, A., Hanakawa, T., Noppeney, U., Devin, J. T., Aso, T., Urayama, S., Fukuyama, H., Stockton, K., Ursui, K., Green, D. W., and Price, C. J. (2006). Language control in the bilingual brain. *Science*, 312:1537–40.
Davis, M. H. and Johnsrude, I. S. (2003). Hierarchical processing in spoken language comprehension. *Journal of Neuroscience*, 23(8):3423–31.
Dehaene, S. (1993). Temporal oscillations in human perception. *Psychological Science*, 4(4):264–70.
Dikker, S., Rabagliati, H., Farmer, T., and Pylkkänen, L. (2010). Early occipital sensitivity to syntactic category is based on form typicality. *Psychological Science*, 21(5):629–34.
Dikker, S., Rabagliati, H., and Pylkkänen, L. (2009). Sensitivity to syntax in visual cortex. *Cognition*, 110(3):293–321.
Ding, N., Melloni, L., Zhang, H., Tian, X., and Poeppel, D. (2016). Cortical tracking of hierarchical linguistic structures in connected speech. *Nature Neuroscience*, 19(1):158–64.
Ding, N. and Simon, J. Z. (2013). Adaptive temporal encoding leads to a background-insensitive cortical representation of speech. *Journal of Neuroscience*, 33(13):5728–35.
Dronkers, N. F., Plaisant, O., Iba-Zizen, M. T., and Cabanis, E. A. (2007). Paul Broca's historic cases: high resolution MR imaging of the brains of Leborgne and Lelong. *Brain*, 130(Pt 5):1432–41.
Dronkers, N. F., Redfern, B. B., and Knight, R. T. (1999). The neural architecture of language disorders. In Gazzaniga, M. S., editor, *The New Cognitive Neurosciences*. Cambridge, MA: MIT Press.
Dronkers, N. F., Wilkins, D. P., Van Valin, R. D., Redfern, B. B., and Jaeger, J. J. (2004). Lesion analysis of the brain areas involved in language comprehension: towards a new functional anatomy of language. *Cognition*, 92(1–2):145–77.
Edgar, J. C., Khan, S. Y., Blaskey, L., Chow, V. Y., Rey, M., Gaetz, W., Cannon, K. M., Monroe, J. F., Cornew, L., Qasmieh, S., Liu, S., Welsh, J. P., Levy, S. E., and Roberts, T. P. L. (2015). Neuromagnetic oscillations predict evoked-response latency delays and core language deficits in autism spectrum disorders. *Journal of Autism and Developmental Disorders*, 45:395–405.
Ehrenhofer, L., Lau, E., and Phillips, C. (To appear). A possible cure for "N400 blindness" to role reversal anomalies in sentence comprehension. This is an unpublished manuscript available at the following link: http://www.colinphillips.net/wp-content/uploads/2019/08/ehrenhofer_lau_phillips_20190318.pdf

Embick, D. and Poeppel, D. (2015). Towards a computational(ist) neurobiology of language: correlational, integrated, and explanatory neurolinguistics. *Language and Cognitive Neuroscience*, 30(4):357–66.

Emmorey, K. and Ozyurek, A. (2014). Language in our hands: neural underpinnings of sign language and co-speech gesture. In *The Cognitive Neurosciences*, pages 657–66. Cambridge, MA: MIT Press.

Etienne, A., Laroia, T., Weigle, H., Afelin, A., Kelly, S. K., Krishnan, A., and Grover, P. (2020). Novel electrodes for reliable EEG recordings on coarse and curly hair. In *2020 42nd Annual International Conference of the IEEE Engineering in Medicine & Biology Society (EMBC)*, pages 6151–4, Montreal, QC, Canada.

Eulitz, C. and Lahiri, A. (2004). Neurobiological evidence for abstract phonological representations in the mental lexicon during speech recognition. *Journal of Cognitive Neuroscience*, 16(4):577–83.

Federmeier, K. D. and Kutas, M. (1999). A rose by any other name: long-term memory structure and sentence processing. *Journal of Memory and Language*, 41(4):469–95.

Feinberg, D. A., Moeller, S., Smith, S. M., Auerbach, E., Ramanna, S., Gunther, M., Glasser, M. F., Miller, K. L., Ugurbil, K., and Yacoub, E. (2010). Multiplexed echo planar imaging for sub-second whole brain fMRI and fast diffusion imaging. *PLoS One*, 5(12):e15710.

Ferreira, F. and Patson, N. (2007). The "good enough" approach to language comprehension. *Language and Linguistics Compass*, 1(1–2):71–83.

Fiebach, C., Schlesewsky, M., Lohmann, G., von Cramon, D., and Friederici, A. (2005). Revisiting the role of Broca's area in sentence processing: syntactic integration versus syntactic working memory. *Human Brain Mapping*, 24(2):79–91.

Frankland, S. M. and Greene, J. D. (2015). An architecture for encoding sentence meaning in left mid-superior temporal cortex. *Proceedings of the National Academy of Sciences of the United States of America*, 112(37):11732–7.

Friederici, A. D. (2012). The cortical language circuit: from auditory perception to sentence comprehension. *Trends in Cognitive Science*, 16(5):262–8.

Friederici, A. D., Chomsky, N., Berwick, R. C., Moro, A., and Bolhuis, J. J. (2017). Language, mind and brain. *Nature Human Behavior*, 1(10):713–22.

Fromkin, V., Rodman, R., and Hyams, N. M. (2017). *An Introduction to Language*. 11th edn. Boston, MA: Cengage.

Fruchter, J., Stockall, L., and Marantz, A. (2013). MEG masked priming evidence for form-based decomposition of irregular verbs. *Frontiers in Human Neuroscience*, 7:798.

Gage, N. Y. Y. and Hickok, G. (2005). Multiregional cell assemblies, temporal binding and the representation of conceptual knowledge in cortex: a modern theory by a "classical" neurologist, Carl Wernicke. *Cortex*, 41(6):823–32.

Gagnepain, P., Henson, R. N., and Davis, M. H. (2012). Temporal predictive codes for spoken words in auditory cortex. *Current Biology*, 22(7):615–21.

Ganong, 3rd, W. F. (1980). Phonetic categorization in auditory word perception. *Journal of Experimental Psychology: Human Perception and Performance*, 6(1):110–25.

Geschwind, N. (1970). The organization of language and the brain. *Science*, 170(3961):940–44.
Geschwind, N. (1972). Language and the brain. *Scientific American*, 226(4):76–83.
Giraud, A.-L., Kleinschmidt, A., Poeppel, D., Lund, T. E., Frackowiak, R. S. J., and Laufs, H. (2007). Endogenous cortical rhythms determine cerebral specialization for speech perception and production. *Neuron*, 56(6):1127–34.
Giraud, A.-L. and Poeppel, D. (2012a). Cortical oscillations and speech processing: emerging computational principles and operations. *Nature Neuroscience*, 15(4):511–17.
Giraud, A.-L. and Poeppel, D. (2012b). Speech perception from a neurophysiological perspective. In Poeppel, D., Overath, T., Popper, A. N., and Fay, R. R., editors, *The Human Auditory Cortex*, pages 225–60. New York: Springer.
Gorno-Tempini, M. L., Dronkers, N. F., Rankin, K. P., Ogar, J. M., Phengrasamy, L., Rosen, H. J., Johnson, J. K., Weiner, M. W., and Miller, B. I. (2004). Cognition and anatomy in three variants of primary progressive aphasia. *Annals of Neurology*, 55(3):335–46.
Gould, S. J. (1981). *The Mismeasure of Man*. New York: Norton.
Gouvea, A. C., Phillips, C., Kazanina, N., and Poeppel, D. (2010). The linguistic processes underlying the P600. *Language and Cognitive Processes*, 25(2):149–88.
Grodzinsky, Y. (2000). The neurology of syntax: language use without Broca's area. *Behavioral and Brain Sciences*, 23(01):1–21.
Grodzinsky, Y., Pinango, M. M., Zurif, E., and Drai, D. (1999). The critical role of group studies in neuropsychology: comprehension regularities in Broca's aphasia. *Brain and Language*, 67(2):134–47.
Grodzinsky, Y. and Santi, A. (2008). The battle for Broca's region. *Trends in Cognitive Sciences*, 12(12):474–80.
Groen, W. B., Zwiers, M. P., van der Gaag, R.-J., and Buitelaar, J. K. (2008). The phenotype and neural correlates of language in autism: an integrative review. *Neuroscience and Biobehavioral Reviews*, 32(8):1416–25.
Grothe, B., Pecka, M., and McAlpine, D. (2010). Mechanisms of sound localization in mammals. *Physiological Reviews*, 90(3):983–1012.
Günther, F., Dudschig, C., and Kaup, B. (2016). Latent semantic analysis cosines as a cognitive similarity measure: evidence from priming studies. *Quarterly Journal of Experimental Psychology*, 69(4):626–53.
Gwilliams, L. and Marantz, A. (2015). Non-linear processing of a linear speech stream: the influence of morphological structure on the recognition of spoken Arabic words. *Brain and Language*, 147(0):1–13.
Hallett, M. (2007). Transcranial magnetic stimulation: a primer. *Neuron*, 55(2):187–99.
Hämäläinen, M., Hari, R., Ilmoniemi, R. J., Knuutila, J., and Lounasmaa, O. V. (1993). Magnetoencephalography: theory, instrumentation, and applications to noninvasive studies of the working human brain. *Reviews of Modern Physics*, 65(2):413–97.

Hartwigsen, G. and Saur, D. (2019). Neuroimaging of stroke recovery from aphasia: insights Insights into plasticity of the human language network. *NeuroImage*, 190:14–31.

Hauk, O., Davis, M. H., Ford, M., Pulvermuller, F., and Marslen-Wilson, W. D. (2006). The time course of visual word recognition as revealed by linear regression analysis of ERP data. *NeuroImage*, 30(4):1383–400.

Haxby, J. V., Connolly, A. C., and Guntupalli, J. S. (2014). Decoding neural representational spaces using multivariate pattern analysis. *Annual Review of Neuroscience*, 37:435–56.

Hickok, G. (2014). *The Myth of Mirror Neurons: The Real Neuroscience of Communication and Cognition*. New York: Norton.

Hickok, G., Okada, K., Barr, W., Pa, J., Rogalsky, C., Donnelly, K., Barde, L., and Grant, A. (2008). Bilateral capacity for speech sound processing in auditory comprehension: evidence from Wada procedures. *Brain and Language*, 107(3):179–84.

Hickok, G. and Poeppel, D. (2007). The cortical organization of speech processing. *Nature Reviews Neuroscience*, 8(5):393–402.

Hickok, G. and Sinigaglia, C. (2013). Clarifying the role of the mirror system. *Neuroscience Letters*, 12:62–6.

Hillis, A. E., Rorden, C., and Fridriksson, J. (2017). Brain regions essential for word comprehension: drawing inferences from patients. *Annals of Neurology*, 81(6):759–68.

Holcomb, P. J. (1988). Automatic and attentional processing: an event-related brain potential analysis of semantic priming. *Brain and Language*, 35(1):66–85.

Holcomb, P. J. and McPherson, W. B. (1994). Event-related brain potentials reflect semantic priming in an object decision task. *Brain and Cognition*, 24(2):259–76.

Hornstein, N. (1999). Movement and control. *Linguistic Inquiry*, 30(1):69–96.

Humphries, C., Binder, J. R., Medler, D. A., and Liebenthal, E. (2006). Syntactic and semantic modulation of neural activity during auditory sentence comprehension. *Journal of Cognitive Neuroscience*, 18(4):665–79.

Huth, A. G., de Heer, W. A., Griffiths, T. L., Theunissen, F. E., and Gallant, J. L. (2016). Natural speech reveals the semantic maps that tile human cerebral cortex. *Nature*, 532(7600):453–8.

Jochaut, D., Lehongre, K., Saitovitch, A., Devauchelle, A.-D., Olasagasti, I., Chabane, N., Zilbovicius, M., and Giraud, A.-L. (2015). Atypical coordination of cortical oscillations in response to speech in autism. *Frontiers in Human Neuroscience*, 9:171.

Jonas, E. and Kording, K. P. (2017). Could a neuroscientist understand a microprocessor? *PLoS Computational Biology*, 13(1):e1005268.

Kelley, E. (2011). Language in ASD. In Fein, D. A., editor, *The Neuropsychology of Autism*. Oxford: Oxford University Press.

Kemmerer, D., Miller, L., Macpherson, M. K., Huber, J., and Tranel, D. (2013). An investigation of semantic similarity judgments about action and non-action verbs in Parkinson's disease: implications for the Embodied Cognition Framework. *Frontiers in Human Neuroscience*, 7:146.

Kenet, T. (2011). Sensory functions in ASD. In Fein, D. A., editor, *The Neuropsychology of Autism*, pages 215–24. Oxford: Oxford University Press.

Kim, A. and Osterhout, L. (2005). The independence of combinatory semantic processing: evidence from event-related potentials. *Journal of Memory and Language*, 52:205–25.

Klimovich-Gray, A., Tyler, L. K., Randall, B., Kocagoncu, E., Devereux, B., and Marslen-Wilson, W. D. (2019). Balancing prediction and sensory input in speech comprehension: the spatiotemporal dynamics of word recognition in context. *Journal of Neuroscience*, 39(3):519–27.

Knudsen, E. I. (2002). Instructed learning in the auditory localization pathway of the barn owl. *Nature*, 417(6886):322–8.

Kravitz, D. J., Saleem, K. S., Baker, C. I., Ungerleider, L. G., and Mishkin, M. (2013). The ventral visual pathway: an expanded neural framework for the processing of object quality. *Trends in Cognitive Sciences*, 17(1):26–49.

Kubanek, J., Brunner, P., Gunduz, A., Poeppel, D., and Schalk, G. (2013). The tracking of speech envelope in the human cortex. *PLoS One*, 8(1):e53398.

Kujala, T., Lepistö, T., and Näätänen, R. (2013). The neural basis of aberrant speech and audition in autism spectrum disorders. *Neuroscience and Biobehavioral Reviews*, 37(4):697–704.

Kuperberg, G. R. (2007). Neural mechanisms of language comprehension: challenges to syntax. *Brain Research*, 1146:23–49.

Kutas, M. and Federmeier, K. D. (2000). Electrophysiology reveals semantic memory use in language comprehension. *Trends in Cognitive Sciences*, 4:463–9.

Kutas, M. and Federmeier, K. D. (2011). Thirty years and counting: finding meaning in the N400 component of the event-related brain potential (ERP). *Annual Review of Psychology*, 62:621–47.

Kutas, M. and Hillyard, S. A. (1980). Reading senseless sentences: brain potentials reflect semantic incongruity. *Science*, 207:203–5.

Kutas, M. and Hillyard, S. A. (1984). Brain potentials during reading reflect word expectancy and semantic association. *Nature*, 307(5947):161–2.

Lago, S., Scharinger, M., Kronrod, Y., and Idsardi, W. J. (2015). Categorical effects in fricative perception are reflected in cortical source information. *Brain and Language*, 143:52–8.

Lahiri, A. and Marslen-Wilson, W. (1991). The mental representation of lexical form: a phonological approach to the recognition lexicon. *Cognition*, 38(3):245–94.

Lambert, J., Eustache, F., Lechevalier, B., Rossa, Y., and Viader, F. (1989). Auditory agnosia with relative sparing of speech perception. *Cortex*, 25(1):71–82.

Laszlo, S. and Federmeier, K. D. (2009). A beautiful day in the neighborhood: an event-related potential study of lexical relationships and prediction in context. *Journal of Memory and Language*, 61(3):326–38.

Lau, E. F., Namyst, A., Fogel, A., and Delgado, T. (2016). A direct comparison of N400 effects of predictability and incongruity in adjective–noun combination. *Collabra*, 2(1):13.

Lau, E. F., Phillips, C., and Poeppel, D. (2008). A cortical network for semantics: (de)constructing the N400. *Nature Reviews Neuroscience*, 9(12):920–33.

Lau, E. F., Stroud, C., Plesch, S., and Phillips, C. (2006). The role of structural prediction in rapid syntactic analysis. *Brain and Language*, 98(1):74–88.

Leonard, M. K., Ferjan Ramirez, N., Torres, C., Travis, K. E., Hatrak, M., Mayberry, R. I., and Halgren, E. (2012). Signed words in the congenitally deaf evoke typical late lexicosemantic responses with no early visual responses in left superior temporal cortex. *Journal of Neuroscience*, 32(28):9700–5.

Lewis, G. and Poeppel, D. (2014). The role of visual representations during the lexical access of spoken words. *Brain and Language*, 134(0):1–10.

Logothetis, N. K. and Wandell, B. A. (2004). Interpreting the BOLD signal. *Annual Review of Physiology*, 66:735–69.

Lopopolo, A., Frank, S. L., van den Bosch, A., and Willems, R. M. (2017). Using stochastic language models (SLM) to map lexical, syntactic, and phonological information processing in the brain. *PLoS One*, 12(5):e0177794.

Luck, S. J. (2014). *An Introduction to the Event-Related Potential Technique*. 2nd edn. Cambridge MA: MIT Press.

MacGregor, L. J., Pulvermüller, F., van Casteren, M., and Shtyrov, Y. (2012). Ultra-rapid access to words in the brain. *Nature Communications*, 3, 711. https://doi.org/10.1038/ncomms1715

MacSweeney, M., Capek, C. M., Campbell, R., and Woll, B. (2008). The signing brain: the neurobiology of sign language. *Trends in Cognitive Sciences*, 12(11):432–40.

Maenner, M. J. (2020). Prevalence of Autism Spectrum Disorder among children aged 8 years. Autism and Developmental Disabilities monitoring Network, USA.

Maffei, C., Capasso, R., Cazzolli, G., Colosimo, C., Dell'Acqua, F., Piludu, F., Catani, M., and Miceli, G. (2017). Pure word deafness following left temporal damage: behavioral and neuroanatomical evidence from a new case. *Cortex*, 97:240–54.

Mahon, B. Z. and Caramazza, A. (2005). The orchestration of the sensory-motor systems: clues from neuropsychology. *Cognitive Neuropsychology*, 22(3):480–94.

Mahon, B. Z. and Caramazza, A. (2008). A critical look at the embodied cognition hypothesis and a new proposal for grounding conceptual content. *Journal of Physiology (Paris)*, 102(1):59–70.

Marinkovic, K., Dhond, R. P., Dale, A. M., Glessner, M., Carr, V., and Halgren, E. (2003). Spatiotemporal dynamics of modality-specific and supramodal word processing. *Neuron*, 38(3):487–97.

Marler, P. (1991). The instinct to learn. In *The Epigenesis of Mind: Essays on Biology and Cognition*, Carey, S. and Gelman, R., editors. Lawrence Erlbaum Associates, pp. 37–66.

Marler, P. (2008). The instinct to learn. In *Brain Development and Cognition*, chapter 17, pages 305–29. Chichester: Wiley.

Marr, D. (1982). *Vision: A Computational Investigation into the Human Representation and Processing of Visual Information*. New York: Freeman.

Matchin, W., Brodbeck, C., Hammerly, C., and Lau, E. (2019). The temporal dynamics of structure and content in sentence comprehension: evidence from fMRI-constrained MEG. *Human Brain Mapping*, 40, 663–78. https://doi.org/10.1002/hbm.24403

Matchin, W., Sprouse, J., and Hickok, G. (2014). A structural distance effect for backward anaphora in Broca's area: an fMRI study. *Brain and Language*, 138:1–11.

Mazoyer, B. M., Tzourio, N., Frak, V., Syrota, A., Murayama, N., Levrier, O., Salamon, G., Dehaene, S., Cohen, L., and Mehler, J. (1993). The cortical representation of speech. *Journal of Cognitive Neuroscience*, 5(4):467–79.

McGurk, H. and MacDonald, J. (1976). Hearing lips and seeing voices. *Nature*, 264(5588):746–8.

Mesgarani, N., Cheung, C., Johnson, K., and Chang, E. F. (2014). Phonetic feature encoding in human superior temporal gyrus. *Science*, 343(6174):1006–10.

Meyer, D. E. and Schvaneveldt, R. W. (1971). Facilitation in recognizing pairs of words: evidence of a dependence between retrieval operations. *Journal of Experimental Psychology*, 90(2):227–34.

Meyer, L. (2018). The neural oscillations of speech processing and language comprehension: state of the art and emerging mechanisms. *European Journal of Neuroscience*, 48(7):2609–21.

Meyer, L., Sun, Y., and Martin, A. E. (2019). Synchronous, but not entrained: exogenous and endogenous cortical rhythms of speech and language processing. *Language, Cognition and Neuroscience*, 35(9):1–11.

Milberg, W. and Blumstein, S. E. (1981). Lexical decision and aphasia: evidence for semantic processing. *Brain and Language*, 14(2):371–85.

Miller, G. A. (2003). The cognitive revolution: a historical perspective. *Trends in Cognitive Sciences*, 7(3):141–4.

Mitchell, T. M., Shinkareva, S. V., Carlson, A., Chang, K.-M., Malave, V. L., Mason, R. A., and Just, M. A. (2008). Predicting human brain activity associated with the meanings of nouns. *Science*, 320(5880):1191–5.

Momma, S. (2016). *Parsing, Generation, and Grammar*. PhD thesis, University of Maryland.

Morillon, B., Liégeois-Chauvel, C., Arnal, L. H., Bénar, C.-G., and Giraud, A.-L. (2012). Asymmetric function of theta and gamma activity in syllable processing: an intra-cortical study. *Frontiers in Psychology*, 3:248.

Möttönen, R. and Watkins, K. E. (2009). Motor representations of articulators contribute to categorical perception of speech sounds. *Journal of Neuroscience*, 29(31):9819–25.

Munson, C. M. (2011). *Perceptual Learning in Speech Reveals Pathways of Processing*. PhD thesis, University of Iowa.

Näätänen, R., Lehtokoski, A., Lennes, M., Cheour, M., Huotilainen, M., Iivonen, A., Vainio, M., Alku, P., Ilmoniemi, R. J., Luuk, A., Allik, J., Sinkkonen, J., and Alho, K. (1997). Language-specific phoneme representations revealed by electric and magnetic brain responses. *Nature*, 385(6615):432–4.

Nelson, M. J., El Karoui, I., Giber, K., Yang, X., Cohen, L., Koopman, H., Cash, S. S., Naccache, L., Hale, J. T., Pallier, C., and Dehaene, S. (2017). Neurophysiological dynamics of phrase-structure building during sentence processing. *Proceedings of the National Academy of Sciences of the United States of America*, 114(18):E3669–78.

Neville, H., Nicol, J. L., Barss, A., Forster, K. I., and Garrett, M. F. (1991). Syntactically based sentence processing classes: evidence from event-related brain potentials. *Journal of Cognitive Neuroscience*, 3(2):151–65.

Nieuwland, M. S., Barr, D. J., Bartolozzi, F., Busch-Moreno, S., Darley, E., Donaldson, D. I., Ferguson, H. J., Fu, X., Heyselaar, E., Huettig, F., Husband, E. M., Ito, A., Kazanina, N., Kogan, V., Kohút, Z., Kulakova, E., Mézière, D., Politzer-Ahles, S., Rousselet, G., Rueschemeyer, S.-A., Segaert, K., Tuomainen, J., and Von Grebmer Zu Wolfsthurn, S. (2019). Dissociable effects of prediction and integration during language comprehension: Evidence from a large-scale study using brain potentials. *Philosophical Transactions of the Royal Society B*. Volume (Issue): 375 (1791)

Nourski, K. V. and Brugge, J. F. (2011). Representation of temporal sound features in the human auditory cortex. *Reviews in the Neurosciences*, 22(2):187–203.

Ojemann, G., Ojemann, J., Lettich, E., and Berger, M. (1989). Cortical language localization in left, dominant hemisphere. An electrical stimulation mapping investigation in 117 patients. *Journal of Neurosurgery*, 71(3):316–26.

Osterhout, L. and Holcomb, P. J. (1992). Event-related potentials elicited by syntactic anomaly. *Journal of Memory and Language*, 31:785–806.

Pallier, C., Devauchelle, A.-D., and Dehaene, S. (2011). Cortical representation of the constituent structure of sentences. *Proceedings of the National Academy of Sciences of the United States of America*, 108(6):2522–7.

Parker Jones, O., Alfaro-Almagro, F., and Jbabdi, S. (2018). An empirical, 21st century evaluation of phrenology. *Cortex*, 106:26–35.

Patterson, K. and Lambon Ralph, M. A. (2016). The hub-and-spoke hypothesis of semantic memory. In *Neurobiology of Language*, pages 765–75. New York: Elsevier.

Patterson, K., Nestor, P. J., and Rogers, T. T. (2007). Where do you know what you know? The representation of semantic knowledge in the human brain. *Nature Reviews Neuroscience*, 8(12):976–87.

Peña, J. L. and DeBello, W. M. (2010). Auditory processing, plasticity, and learning in the Barn Owl. *ILAR Journal*, 51(4):338–52.

Phillips, C. (2006). The real-time status of island phenomena. *Language*, 82(4):795–823.

Phillips, C., Pellathy, T., Marantz, A., Yellin, E., Wexler, K., Poeppel, D., McGinnis, M., and Roberts, T. (2000). Auditory cortex accesses phonological categories: an MEG mismatch study. *Journal of Cognitive Neuroscience*, 26(6):1038–55.

Pinker, S. and Ullman, M. T. (2002). The past and future of the past tense. *Trends in Cognitive Sciences*, 6(11):456–63.

Pobric, G., Jefferies, E., and Ralph, M. A. L. (2007). Anterior temporal lobes mediate semantic representation mimicking semantic dementia by using rTMS in normal participants. *Proceedings of the National Academy of Sciences of the United States of America*, 104(50), 20137–41. https://doi.org/10.1073/pnas.0707383104

Poeppel, D. (2001). Pure word deafness and the bilateral processing of the speech code. *Cognitive Science*, 25(5):679–93.

Poeppel, D. (2003). The analysis of speech in different temporal integration windows: cerebral lateralization as "asymmetric sampling in time." *Speech Communication*, 41(1):245–55.

Poeppel, D. and Embick, D. (2005). Defining the relation betweein linguistics and neuroscience. In Cutler, A., editor, *Twenty-First Century Psycholinguistics: Four Cornerstones*, chapter 6. Mahwah, NJ: Erlbaum.

Poldrack, R. A., Mumford, J. A., and Nichols, T. E. (2011). *Handbook of Functional MRI Data Analysis*. Cambridge: Cambridge University Press.

Polster, M. R. and Rose, S. B. (1998). Disorders of auditory processing: evidence for modularity in audition. *Cortex*, 34:47–65.

Pulvermüller, F., Hauk, O., Nikulin, V. V., and Ilmoniemi, R. J. (2005). Functional links between motor and language systems. *European Journal of Neuroscience*, 21(3):793–7.

Pulvermüller, F., Huss, M., Kherif, F., Moscoso del Prado Martin, F., Hauk, O., and Shtyrov, Y. (2006). Motor cortex maps articulatory features of speech sounds. *Proceedings of the National Academy of Sciences of the United States of America*, 103(20):7865–70.

Pylkkänen, L. (2016). Composition of complex meaning: interdisciplinary perspectives on the left anterior temporal lobe. In Hickok, G. and Small, S., editors, *Neurobiology of Language*. London: Academic Press.

Pylkkänen, L., Martin, A. E., McElree, B., and Smart, A. (2008). The anterior midline field: coercion or decision making? *Brain and Language*, 108(3):184–90.

Ramón y Cajal, S. (1899). *Comparative Study of the Sensory Areas of the Human Cortex*. Worcester, MA: Clark University.

Rapin, I., Dunn, M. A., Dunn, M. A., Allen, D. A., Stevens, M. C., and Fein, D. (2009). Subtypes of language disorders in school-age children with autism. *Developmental Neuropsychology*, 34(1):66–84.

Ringach, D. L. (2004). Mapping receptive fields in primary visual cortex. *Journal of Physiology*, 558(3):717–28.

Rizzolatti, G. and Arbib, M. A. (1998). Language within our grasp. *Trends in Neuroscience*, 21(5):188–94.

Roberts, T. P. L., Ferrari, P., Stufflebeam, S. M., and Poeppel, D. (2000). Latency of the auditory evoked neuromagnetic field components: stimulus dependence and insights toward perception. *Journal of Clinical Neurophysiology*, 17(2):114–29.

Roberts, T. P. L., Khan, S. Y., Rey, M., Monroe, J. F., Cannon, K., Blaskey, L., Woldoff, S., Qasmieh, S., Gandal, M., Schmidt, G. L., Zarnow, D. M., Levy, S. E., and Edgar, J. C. (2010). MEG detection of delayed auditory evoked responses in autism spectrum disorders: towards an imaging biomarker for autism. *Autism Research*, 3(1):8–18.

Rogalsky, C. and Hickok, G. (2010). The role of Broca's area in sentence comprehension. *Journal of Cognitive Neuroscience*, 23(7):1–17.

Rogalsky, C., Love, T., Driscoll, D., Anderson, S. W., and Hickok, G. (2011). Are mirror neurons the basis of speech perception? Evidence from five cases with damage to the purported human mirror system. *Neurocase*, 17(2):178–87.

Rogers, T. T., Lambon Ralph, M. A., Garrard, P., Bozeat, S., McClelland, J. L., Hodges, J. R., and Patterson, K. (2004). Structure and deterioration of semantic memory: a neuropsychological and computational investigation. *Psychological Review*, 111(1):205–35.

Rossi, S., Telkemeyer, S., Wartenburger, I., and Obrig, H. (2012). Shedding light on words and sentences: near-infrared spectroscopy in language research. *Brain and Language*, 121(2):152–63.

Rumiati, R. I., Zanini, S., Vorano, L., and Shallice, T. (2001). A form of ideational apraxia as a delective deficit of contention scheduling. *Cognitive Neuropsychology*, 18(7):617–42.

Saberi, K. and Perrott, D. R. (1999). Cognitive restoration of reversed speech. *Nature*, 398(6730):760.

Sagan, C. (1979). *Broca's Brain: Reflections on the Romance of Science*. New York: Random House.

Samson, D. and Pillon, A. (2003). A case of impaired knowledge for fruit and vegetables. *Cognitive Neuropsychology*, 20(3):373–400.

Saoud, H., Josse, G., Bertasi, E., Truy, E., Chait, M., and Giraud, A.-L. (2012). Brain-speech alignment enhances auditory cortical responses and speech perception. *Journal of Neuroscience*, 32(1):275–81.

Scharinger, M., Bendixen, A., Trujillo-Barreto, N. J., and Obleser, J. (2012). A sparse neural code for some speech sounds but not for others. *PLOS ONE*, 7(7):e40953.

Scharinger, M., Idsardi, W. J., and Poe, S. (2011). A comprehensive three-dimensional cortical map of vowel space. *Journal of Cognitive Neuroscience*, 23(12):3972–82.

Shannon, R. V., Zeng, F. G., Kamath, V., Wygonski, J., and Ekelid, M. (1995). Speech recognition with primarily temporal cues. *Science*, 270(5234):303–4.

Signoret, J.-L., Castaigne, P., Lhermitte, F., Abelanet, R., and Lavorel, P. (1984). Rediscovery of Leborgne's brain: anatomical description with CT scan. *Brain and Language*, 22(2):303–19.

Sohoglu, E., Peelle, J. E., Carlyon, R. P., and Davis, M. H. (2012). Predictive top-down integration of prior knowledge during speech perception. *Journal of Neuroscience*, 32(25):8443–53.

Solomyak, O. and Marantz, A. (2009). Lexical access in early stages of visual word processing: a single-trial correlational MEG study of heteronym recognition. *Brain and Language*, 108(3):191–6.

Sporns, O. (2016). *Networks of the Brain*. Cambridge, MA: MIT Press.

Steinhauer, K. and Drury, J. E. (2012). On the early left-anterior negativity (ELAN) in syntax studies. *Brain and Language*, 120(2):135–62.

Stowe, L. A. (1986). Parsing WH-constructions: evidence for on-line gap location. *Language and Cognitive Processes*, 1(3):227–45.

Stowe, L. A., Broere, C. A., Paans, A. M., Wijers, A. A., Mulder, G., Vaalburg, W., and Zwarts, F. (1998). Localizing components of a complex task: sentence processing and working memory. *Neuroreport*, 9(13):2995–9.

Talavage, T. M., Sereno, M. I., Melcher, J. R., Ledden, P. J., Rosen, B. R., and Dale, A. M. (2004). Tonotopic organization in human auditory cortex revealed by progressions of frequency sensitivity. *Journal of Neurophysiology*, 91(3):1282–96.

Uttal, W. R. (2003). *The New Phrenology: The Limits of Localizing Cognitive Processes in the Brain*. Cambridge, MA: MIT Press.

Van Berkum, J. J. A., van den Brink, D., Tesink, C. M. J. Y., Kos, M., and Hagoort, P. (2008). The neural integration of speaker and message. *Journal of Cognitive Neuroscience*, 20(4):580–91.

Van Berkum, J. J. A., Zwitserlood, P., Hagoort, P., and Brown, C. M. (2003). When and how do listeners relate a sentence to the wider discourse? Evidence from the N400 effect. *Journal of Cognitive Neuroscience*, 17(3):701–18.

Van Petten, C. and Kutas, M. (1990). Interactions between sentence context and word frequency in event-related brain potentials. *Memory & Cognition*, 18(4):380–93.

Van Petten, C. and Luka, B. J. (2012). Prediction during language comprehension: benefits, costs, and ERP components. *International Journal of Psychophysiology*, 83(2):176–90.

van Wassenhove, V., Grant, K. W., and Poeppel, D. (2007). Temporal window of integration in auditory-visual speech perception. *Neuropsychologia*, 45(3):598–607.

von Frisch, K. (1974). Decoding the language of the bee. *Science*, 185(4152):663–8.

Warrington, E. K. and McCarthy, R. (1983). Category specific access dysphasia. *Brain*, 106(4): 859–78.

Warrington, E. K. and Shallice, T. (1984). Category specific semantic impairments. *Brain*, 107(3):829–54.

Wehbe, L., Murphy, B., Talukdar, P., Fyshe, A., Ramdas, A., and Mitchell, T. (2014). Simultaneously uncovering the patterns of brain regions involved in different story reading subprocesses. *PLoS One*, 9(11):e112575.

Wernicke, C. (1874). *Der aphasische Symptomencomplex: Eine psychologische Studie auf anatomischer Basis*. Breslau: Cohn & Weigert.

Wernicke, C. (1977). Der aphasische Symptomkomplex: Eine psychologische Studie auf anatomischer Basis. In Eggert, G. H., editor, *Wernicke's Works on Aphasia: A Sourcebook and Review*. The Hague: Mouton.

Wilson, S. M., DeMarco, A. T., Henry, M. L., Gesierich, B., Babiak, M., Mandelli, M. L., Miller, B. L., and Gorno-Tempini, M. L. (2014). What role does the anterior temporal lobe play in sentence-level processing? Neural correlates of syntactic processing in semantic variant primary progressive aphasia. *Journal of Cognitive Neuroscience*, 26(5):970–85.

Wilson, S. M., Galantucci, S., Tartaglia, M. C., and Gorno-Tempini, M. L. (2012). The neural basis of syntactic deficits in primary progressive aphasia. *Brain and Language*, 122(3):190–8.

Yi, H. G., Leonard, M. K., and Chang, E. F. (2019). The encoding of speech sounds in the superior temporal gyrus. *Neuron*, 102(6):1096–110.

Zaccarella, E. and Friederici, A. D. (2015). Merge in the human brain: a sub-region based functional investigation in the left pars opercularis. *Frontiers in Psychology*, 6:1818.

Zaccarella, E., Meyer, L., Makuuchi, M., and Friederici, A. D. (2017a). Building by syntax: the neural basis of minimal linguistic structures. *Cerebral Cortex*, 27(1):411–21.

Zaccarella, E., Schell, M., and Friederici, A. D. (2017b). Reviewing the functional basis of the syntactic Merge mechanism for language: a coordinate-based activation likelihood estimation meta-analysis. *Neuroscience and Biobehavioral Reviews*, 80:646–56. https://doi.org/10.1016/j.neubiorev.2017.06.011

Zhang, L. and Pylkkänen, L. (2015). The interplay of composition and concept specificity in the left anterior temporal lobe: an MEG study. *Neuroimage*, 111: 228–40.

Zurif, E. B. and Pinango, M. M. (1999). The existence of comprehension patterns in Broca's aphasia. *Brain and Language*, 70(1):133–8.

Index

intracranial electrodes (ICE), *see* electrocorticography

action potential, 24, 44, 45
agnosia, 29, 73, 101, 175
　auditory, 73
　category-specific, 101, 102
agrammaticism, 95, 96, 162
agreement, 117, 162
algorithmic level, *see* levels of description, 16
alpha waves, 56, *see also* neural oscillations
Alzheimer's Disease, 106
analysis by synthesis, 59–61, 74, 76, 80
anaphora, 161, 175, 184
angular gyrus, 22, 102
anomia, 95, 96, 175
anterior temporal lobe, 86, 87, 89, 105–110, 115, 136, 142, 146, 158
　left, 108, 109, 144–148, 150, 152, 154, 155, 163, 164, 184
aphasia, 13, 14, 16, 17, 27, 31, 83, 95, 97, 98, 106, 162, 164, 175
　fluent, 12–14, 17, 28, 30, 85
　non-fluent, 10, 13, 14, 16, 27, 28, 30, 75, 95, 156, 157
apraxia, 113, 114
arcuate fasciculus, 14, 142
argument structure, 117, 129, 149, 152–155, 163
ASD, *see* autism spectrum disorder
ATL, *see* anterior temporal lobe
auditory cortex, 45–51, 53–56, 60, 63, 64, 74, 78–80, 86, 87, 89, 92, 93, 136, 137
Autism Spectrum Disorder, 81

backwards anaphora, 161, 162
barn owl, 5–7
bigram, 92
bilingual, 108, 109
block design, 32–34, 152
blood oxygentation-level dependent signal, 32–34, 46, 49, 56, 167, 168, 180
BOLD signal, *see* blood oxygentation-level dependent
Broca's aphasia, *see* aphasia, non-fluent
Broca's area, *see* inferior frontal gyrus, left
Brodmann's areas, 23, 25

categorical perception, 52–54, 72, 181
central sulcus, 20, 21, 25
classical model, 11–14, 17, 30, 156, 175
Cloze task, 123, 124
cochlea, 44–46, 60, 90
cocktail party effect, 48
coincidence detector, 5, 6
coindidence detector, 5
combinatoric network, 140–143, 148, 154, 165
compositionality, 117, 137, 148, 163, 165
compositionally, 118
computational level, *see* levels of description, 16
conceptual specificity, 105, 145, 147, 155
conditional probability, 135, 136
constituency, 116–118, 127, 128, 137, 139, 142, 146, 153, 155, 164, 168
cytoarchitecture, 23, 24, 175

DCS, *see* direct cortical stimulation
deficit/lesion method, 27, 28, 30, 39, 40, 83, 92, 95, 98, 156, 159, 184, 204
deficit/lesion method, 92
delay line, 5, 6
delayed-copy task, 105, 106
dichotic listening, 57
diffusion tensor imaging, 26, 142, 175
direct cortical stimulation, 39, 180
distributed-only theory, 105
distributed-plus-hub, 183
distributed-plus-hub theory, 105, 107
domain-general, 159, 161, 162, 164
domain-specific, 159, 161, 162, 164, 165
dorsal stream, 62, 63
double dissociation, 28–30, 73, 95–98, 113, 114
DTI, *see* diffusion tensor imaging
dual-stream model, 62, 63, 86

early left anterior negativity, 183
ECoG, *see* electrocorticography
EEG, *see* electroencephalography
ELAN, *see* early left anterior negativity
elecetroencelphalography, 35–38, 56, 69, 79, 119, 124, 127, 133, 167, 180, 181, 183
electrocorticography, 38, 39, 48, 53, 54, 66, 68, 69, 175
embodied concepts, 101, 110–115
entrainment, 48
episodic memory, 106
ERF, *see* event-related field
ERP, *see* event-related potential
event-related design, 32, 33
event-related field, 38
event-related potential, 35–37, 71, 119–121, 124, 127, 130–135, 138, 168, 175, 180, 183

field mouse, 6, 7
filled-gap effect, 184
fine structure, 46, 47, 55–57, 61, 76, 79, 80, 86

fMRI, *see* functional magnetic resonance imaging
fMRI adaptation, 108
fNIRS, *see* functional near-infrared spectroscopy
full decomposition theory, 94, 95, 97–99
functional magnetic resonance imaging, 31–34, 46, 49, 56, 67, 68, 74, 79, 86, 102–104, 108, 110, 134, 136, 140, 149–151, 153, 154, 161, 162, 164, 166, 170, 171, 175, 180, 184
functional near-infrared spectroscopy, 34
fusiform gyrus, 22

Ganong effect, 52, 54
grammar, 8, 15, 94, 118, 119, 129, 160, 162
grammatical role, 149, 157
grounded symbolic concepts, 110, 112, 114, 115

hemodynamic response function, 32, 33, 180
Heschl's gyrus, *see* auditory cortex
hierarchical structure, *see* constituency

IFG, *see* inferior frontal gyrus
implementational level, *see* levels of description, 16
inferior frontal gyrus, 22, 25, 28, 59, 87, 167, 169
 left, 13, 28, 40, 136, 140, 142, 144, 149, 154–156, 158–165, 184
inferior frontal lobe, 89
inferior parietal lobe, 108, 111
inferior temporal gyrus, 21, 22, 148
intracranial electrodes (ICE), *see* eectrocorticography
IPL, *see* inferior parietal lobe

jabberwocky, *see* pseudoword

language dominance, 181
lateral sulcus, 21, 22

LATL, *see* anterior temporal lobe, left
lesion overlap, 31, 83, 158
levels of description, 2, 3, 5, 7, 9, 160, 167
lexical cohort, 92
lexical decision task, 85, 112
lexical item, 62, 82, 83, 85, 86, 90, 92–95, 97–100, 106, 109, 114, 130, 133, 134, 136, 156
lexical semantics, 101, 104
LIFG, *see* inferior frontal gyrus, left
linguistic competence, 180
linguistic performance, 180
linking function, *see* linking hypothesis
linking hypothesis, 9, 16, 119, 143, 160, 167, 176
localizationist, 10, 12, 14, 170, 171, 179
long-distance dependency, 117, 139, 157–162, 164, 168

M100, 64, 78, 81, 176
magnetic resonance imaging, 25–28, 31–33, 37, 158, 175, 176
magnetoencephalography, 35–38, 49, 64, 66, 69, 78, 86, 87, 89–92, 96–99, 136, 144, 146, 150, 151, 155, 166, 167, 176, 180, 181, 184
McGurk effect, 57, 58, 72
MEG, *see* magnetoencephalography
mental lexicon, 81, 92, 94, 131
meta-analysis, 102, 103, 141
microprocessor, 4, 5
middle temporal gyrus, 21, 148
  posterior, 85, 86, 89, 93
mismatch response, 69–71, 81, 176, 181
MMF, *see* mismatch response
MMN, *see* mismatch response
MMR, *see* mismatch response
motor cortex, 21, 39, 74, 75, 77, 108, 111–114
MRI, *see* magnetic resonance imaging
multi-voxel pattern analysis, 67, 109, 149, 150, 153, 181
MVPA, *see* multi-voxel pattern analysis
myelin, 24

N400, 119–128, 130, 131, 133, 138, 176, 183
naturalistic, 103, 134, 152–154, 184
neural oscillations, 55–57, 72, 76, 78, 79, 163, 171, 176, 177
neurogram, 44, 51–54, 60, 64, 65, 74, 80, 176
noise vocoding, 47, 59, 86, 181
nucleus laminaris, 6

oddball design, 69–71

P600, 131–135, 138, 168, 176
Parkinson's disease, 112, 113
pars opercularis, 164, *see also* inferior frontal gyrus
pars triangularis, 158, 162, 164, *see also* inferior frontal gyrus
parsing, 8, 118, 119, 160, 161, 163
part of speech, *see* syntactic category
partial decomposition theory, 95–98
passive voice, 134, 135, 157, 159, 184
periodotopy, 51, 60, 176
PET, *see* positron emission tomography
phoneme monitoring task, 62
phonemic receptive fields, 53, 66, 68, 69, 81
phonological sketch, 44, 51, 60, 61, 64, 80
phonotopy, 66, 68, 181
phrase structure, *see* constituency
phrenology, 11, 179
picture-matching task, 144, 158
picture-naming task, 29, 72, 75, 76, 84, 85, 101, 106
picture-sequencing task, 113
plausibility judgment, 123
pMTG, *see* posterior middle temporal gyrus
POS, *see* syntactic category
positron emission tomography, 34
posterior middle temporal gyrus, 83, 86, 87, 92, 93, 99, 106, 137, *see also* posterior temporal lobe
posterior temporal lobe, 89, 136

left, 13, 28, 149–156, 163, 164
PPA, *see* primary progressive aphasia
primary progressive aphasia, 106, 115, 158, 159, 162, 176, 184
pseudoword, 90, 91, 102, 140–143, 150, 152, 183
pure word deafness, 72, 73, 77
PWD, *see* pure word deafness

receptive field, 47, 67, 68, 181
reduced relative clause, 132
rTMS, *see* transcranial magnetic stimulation, repeated

semantic combination, 146
semantic composition, 146, 154, 163, *see also* parsing
semantic dementia, 105–108, 115, 146–148, 158, 182
semantic hub, 106, 108–110, 115, 146, 176
semantic integration theory, 122–124
semantic memory, 120
semantic memory theory, 120–124, 133
semantic P600, 134
semantic priming, 85, 104, 108, 109, 120–122, 183
semantic role reversal, 128, 129, 157–159
semantic similarity judgment, 113
sign language, 44, 82, 89, 146, 172, 181, 182
simple composition, 143, 145–147, 149, 153, 164
single dissociation, 29
single-trial analysis, 92, 97, 123
SMG, *see* surpamarginal gyrus
somatosensory cortex, 21, 111
sparse sampling, 49
speech envelope, 46–50, 54, 55, 57, 61, 79, 80
speech is special hypothesis, 60, 72, 77

stereoencephalography (sEEG), *see* electrocorticography
structural connectivity, 26, 142, 165, 175
superior temporal gyrus, 21, 22, 45, 53, 59, 60, 63, 65–68, 76, 86, 87, 92, 93, 99, 148, 154, *see also* anterior temporal lobe, posterior temporal lobe
left, 64, 140
superior temporal sulcus, 92
supramarginal gyrus, 22
sylvian fissure, *see* lateral sulcus
syntactic category, 91, 117, 127, 135, 136, 138
syntactic reanalysis, 134, 138
syntactic structure-building, 7, 146, 147, 151, 155, *see also* parsing

tDCS, *see* transcranial direct current stimulation
temporal pole, 141, 142, 148
temporal windows of integration, 55–58, 76–80, *see also* neural oscillations, 177
temporal-parietal junction, 152, 154
thematic role, 149, 151, 157
TMS, *see* transcranial magnetic stimulation
tonotopy, 46, 47, 49, 50, 60, 66, 177
trace-deletion hypothesis, 159, 161
transcranial direct current stimulation, *see also* direct cortical stimulation
transcranial magnetic stimulation, 39, 40, 107, 110–112, 166, 177, 180
repeated, 74, 75, 77, 107–109, 111

VBM, *see* voxel-based morphometry
ventral stream, 62, 63
ventro-medial prefrontal cortex, 144, 184
visual cortex, 87, 88, 91, 181
visual word-form area, *see* fusiform gyrus

VLSM, *see* voxel-based lesion symptom mapping
VMPFC, *see* ventro-medial prefrontal cortex, 146
voxel-based lesion symptom mapping, 84, *see also* lesion overlap
voxel-based morphometry, 158

Wada procedure, 76, 77
Wernicke's aphasia, *see* aphasia, fluent
Wernicke's area, *see* posterior temporal lobe, left
wh-question, 159, 160–162, 184
word embedding, 103, 104, 136, 177, 182
working memory, 159–164